ROUTLEDGE LIBRARY EDITIONS:
HUMAN GEOGRAPHY

Volume 11

T0264811

THE POLITICS OF LOCATION

THE POLITICS OF LOCATION
An introduction

ANDREW KIRBY

Routledge
Taylor & Francis Group

LONDON AND NEW YORK

First published in 1982 by Methuen & Co. Ltd

This edition first published in 2016
by Routledge
2 Park Square, Milton Park, Abingdon, Oxon OX14 4RN

and by Routledge
711 Third Avenue, New York, NY 10017

Routledge is an imprint of the Taylor & Francis Group, an informa business

British Library Cataloguing in Publication Data
A catalogue record for this book is available from the British Library

ISBN: 978-1-138-95340-6 (Set)
ISBN: 978-1-315-65887-2 (Set) (ebk)
ISBN: 978-1-138-96123-4 (Volume 11) (hbk)
ISBN: 978-1-315-65989-3 (Volume 11) (ebk)

Publisher's Note
The publisher has gone to great lengths to ensure the quality of this reprint but
points out that some imperfections in the original copies may be apparent.

Disclaimer
The publisher has made every effort to trace copyright holders and would welcome
correspondence from those they have been unable to trace.

The politics of location

An introduction

Andrew Kirby

Methuen
London and New York

First published in 1982 by
Methuen & Co. Ltd
11 New Fetter Lane, London EC4P 4EE

Published in the USA by
Methuen & Co.
in association with Methuen, Inc.
733 Third Avenue, New York, NY 10017

© *1982 Andrew Kirby*

Typeset by Keyset Composition,
Colchester and printed in Great Britain at
the University Press, Cambridge

British Library
Cataloguing in Publication Data

Kirby, Andrew
The politics of location.
1. Anthropo-geography   2. Space and
time
I. Title
304.2   GF71

ISBN 0-416-33900-X
ISBN 0-416-33910-7 Pbk

Library of Congress
Cataloging in Publication Data

Kirby, Andrew
The politics of location.
(University paperbacks; UP 786)
Includes index.
1. Space in economics.   2. Geography,
Political.   3.   Anthropo-geography.
4. City planning – Great Britain.
I. Title
HB199.K424 1982   304.2   82-8132

ISBN 0-416-33900-X   AACR2
ISBN 0-416-33910-7 (pbk.)

Dedicated to Alfred Smith Hartley
with love and thanks

Contents

Acknowledgements

I am glad to acknowledge the following individuals and institutions who have willingly, and generally freely, made their work available for reproduction: the Editor of Economic Geography, for permission to use Tables 1 and 4 from *Economic Geography*, April 1980, pp 89–109; Dr R. A. King, Exeter University, for permission to use Table 1 from *Social and Economic Administration*, 1971, 5(3), pp 165–175; the Executive Director of the AAG for permission to use Tables 2 and 6 from the *Annals* of the AAG, 1980, 70(3), pp 342–52; the Editor of the *Journal of Advanced Nursing* 1977, 2, pp 609–19; the Almqvist and Wiksell Periodical Company and the author, Professor J. O. Wheeler, for permission to use material from *Geografiska Annaler* 56B 1976, pp 67–78; Professor R. J. Johnston and the Electoral Reform Society for permission to use Table 1 from *Representation*, 18, 72, pp 23–36; E. J. Brill Boekhandel en Drukkerij for permission to use Table 1 from *Behaviour* 1979, 71 (1–2), pp 146–166; the General Secretary of the Institute for the Study of Treatment of Delinquency for permission to use material from the *British Journal of Criminology* 21(1), 1981, pp 27–46; Dr P. J. Taylor, Newcastle University, for permission to reproduce Figure 1 from his Departmental Seminar Paper 37, 1980; John Bale, University of Keele, for permission to reproduce his Figure from *Town and Country Planning* 49(3), pp 93–4; Professor D. M. Smith of London University for permission to reproduce Figure 2 from *Values, Relevance and Policy*, 1977; Dr Malcolm Moseley, University of East Anglia, for permission to use part of Figure 4.10 from *Accessibility: the rural challenge*; the Institute of British Geographers, for permission to reproduce Figure 7 from 'Cosmetic Planning or Social Engineering', *Area* 1974(4); the Directors of Open Books Publishing Ltd. for permission to use Figure 5.6 from Rutter et al. *15000 Hours*, 1979; the Editor of the *Journal of the Royal College of General Practitioners* for permission to use Figure

2 from the article by Paul Knox, volume 29, 1979, pp 162–168; the Editors of *Urban Studies* and Dr John Ashby of Pion Ltd. for permission to use material from *Urban Studies* 13, 1976, pp 13–25 and *Environment and Planning A*, 8, 1976, pp 43–58, which together make up Figure 4.2; the Punch Library with respect to the Trog cartoon that constitutes Figure 5.1: this is reproduced by permission of *Punch*; Professor J. M. Batty for permission to use Figure 2.1 from his contribution to *Resources and Planning*, 1979; and the Association of Metropolitan Authorities who generously supplied the illustrations incorporated in Figure 7.1. My thanks go to all these sources, and to those whose work I have quoted or from which I have used brief extracts.

Preface

It has been said that 'a poem is never completed, it is only abandoned' (Paul Valery). I would not like to imply that this book has been abandoned, in the sense that it has been tossed aside; none the less, it is certainly not complete. Rather, it represents an introduction to a line of thought; it is an attempt to isolate some particular lines of enquiry, a series of relationships. It is, by necessity, a fairly broad perspective, and it is this which has caused me to stop writing at this particular time. As I argue at rather greater length in Chapter 8, a good deal of the argument that follows is concerned with broad interpretations, and very detailed explanations of social processes have not been attempted. Nevertheless they are critically important, and as a result I have been faced with two options. The first is to continue reforming the arguments presented here, as my own ideas (and the thoughts of others) develop. This is however not likely to be a speedy task, nor one readily accomplished in all the contexts discussed below. In consequence, I have opted for the second course, which is to present my current ideas, and to begin work on a more detailed book, which will concentrate solely upon the field of public-service provision.

The intentions of this volume are straightforward: to present a coherent account of the value of a spatial perspective. As I stress in Chapters 1 and 8 and in the Introduction, this does not imply a unique geography; rather, it explicitly draws upon other ideas, notably those of Weber and Dahrendorf. Nor, as already stated, is it a completely firm edifice with an explanation for everything; reviewers and readers will doubtless comment upon the ahistorical treatment of much of the material, for instance. That kind of historical approach will serve as the ultimate test of the thesis sketched here, and work on the necessarily detailed studies involved is underway.

This book has had its teething problems, and I have been supported by numerous people at Reading; in particular Sheila Dance and Brian Rogers, who drew the diagrams, and Chris Holland, who wore down her fingers – again. Several generations of students have provided critical remarks (thank you, Richard), and slightly older colleagues have provided more formal, but no less critical comments: my thanks go to Kelvyn Jones, Alan Hooper, Peter Hall, John Short, Ron Johnston and Alan Burnett, although the usual disclaimers must apply. They may not detect the value of their remarks, but it *was* useful. Finally, may I thank Mary Ann Kernan, for providing – at the vital moment – consistent help and encouragement; she is all an editor should be.

Andrew Kirby
Cholsey
January 1982

Introduction

'You have intense loyalties to specific places and environments, don't you?'
'Yes, and it is the total environment. A human being within an environment is a reflection of all the aspects of that environment.'
(Thomas Hoving: *Two Worlds of Andrew Wyeth*)

In his book *Clochemerle* Gabriel Chevalier described the way in which an environment can create a unique collection of individuals, shaped by a particular mix of history and geography, chance and the predetermined. Near to Clochemerle were many other villages, but these too were represented as unique. After reading *Clochemerle*, it becomes easy for one to see why a generation of geographers, both French and English, dedicated themselves to the study of the *pays*, that part of the landscape in which the human and the physical meshed together in a distinctive manner.

This book is not an attempt to return to such a tradition; nor was it ever a particularly successful approach. Throughout Chevalier's tale of one French village we are constantly reminded of the world outside: soldiers return from wars, socialists quarrel with clerics, civil servants in Paris make their decisions. Within contemporary society, these external links are even more varied and more immediate. This does not negate, however, a simple, general principle, namely that each individual is located within a specific environment. That environment still possesses its physical aspects: even within a highly unionised industry such as mining in Britain, a miner's bonuses may depend simply upon the thickness of the seams in his pit. Increasingly, however, the individual is affected far more by social issues. To live in a particular location is to face a whole range of phenomena that can change the nature – and quality – of one's life. Employment opportunities vary dramatically from region to region. Life expectancy also varies. Public provision, too, differs in its quality between, and even within, neighbouring towns; transportation, schooling, health care, the library service, even the quality of the roads vary from place to place.

The description of these variations – typically using social indicators – is now well established; the task of explanation is not. Since the mid-1970s the intellectual trend has been to push back the level of explanation, first from individuals and ultimately to the state. As I shall suggest below, this poses some problems for the study of public provision and organization, which is the focus of this book. More immediately, the changing nature of explanation has also produced its own problems.

Initial attempts to account for public-policy variations (in, for instance, housing provision) tended to draw on a weakly articulated managerialism which was often only defined as such once Ray Pahl (1970) had outlined his thoughts on managers and the managed. Unsurprisingly, much of this work appears unfocused in the face of subsequent criticisms, many of which have emphasized structures rather than individual actions and motivations.

The materialist account (the strongest source of structuralist ideas) has had its clear successes (particularly in the context of uneven development), but it has not produced similar breakthroughs in the field of what has come to be known as welfare geography (see, for example, Chapter 2). Eyles has recently observed that:

this grafting on of Marxist concepts to an essentially positivist welfare approach does not work within the constraints of Marxism. To use such concepts in a meaningful way requires a different starting point with qu stions posed in different ways. In welfare geography, the questions are not posed in a dialectical, holistic way. (1981, p. 1375)

In simple terms, welfare geography (or the study of 'well-being' or 'inequality') is based in the descriptive present, and lacks any historical analysis: its specificity is assumed away – a problem touched upon again in Chapter 8.

A further problem revealed by these attempts to utilize materialist concepts is the schizophrenia engendered with respect to space itself. Eyles once more neatly summarizes this problem:

through a strong Marxist theoretical stance . . . questions posed in the domain of geography will become recomposed, redefined and incorporated into a theoretical system which nullifies geography's existence . . . this view sees spatial differentiation as a detail resulting from the operation of social systems. It is not a fundamental property of such systems. (1981, p.1378)

This reading of the problem is superficially correct, but overlooks the varied nature of space. As I argue in Chapter 1, of central importance here is the fact that space need not be reified in philosophical terms, as it is already reified in the context of organizational boundaries. As the examples in Chapter 2 indicate, public provision – be it by the state or the 'local' state – takes place via an organizational framework of explicit spatial structures. It consequently makes solid sense to focus upon variations in provision and consumption between these spatial units, and even the nature of the units themselves.

So far, this argument suggests that a Marxist perspective – particularly a focus upon the state – might still be useful. Here, however, I have emphasized an explicitly Weberian approach, which examines the outcomes of the consumption process. To most materialists, production and consumption are inseparable. However, consumption issues can cross-cut class issues: they have effects upon groups which have only location – not class – in common. Thus, the political issues engendered by consumption conflicts (Chapters 1 and 3; see also 5 and 6), cannot be simply read-off from the ultimately restrictive Marxist account of social relations; nor can the impacts upon individuals in the contexts of health care, education and similar fields in which public provision can determine the quality of life.

Let us be clear however where this argument leads. It does *not* suggest that space is resuscitated as a single focus of analysis. This would clearly be a retrograde step, as it would raise geographical concerns to primacy above an interest in the state: or for others, individuals. Furthermore, there exist many contexts, even in the sphere of consumption, where an aspatial perspective suggests itself: as Saunders argues, for example, various aspects of public provision are organized without regard to geography (1981, pp. 210–11). These notwithstanding, there still remain several contexts within which a spatial perspective is necessary: the drawing of boundaries between local states, and the conflicts that exist between them (Kirby, 1982a); the differential financing of local affairs (Chapter 7); the political struggles developed in relation to the siting of externalities (Part III); and the numerous instances in which space becomes reified by boundary commissioners, education officers, or family practitioner committees, as a result of the complex interactions of the wide varieties of struggle at both the local level and between the state and the 'local' state (Kirby, 1982c).

This book is thus both an attempt to provide a coherent focus for spatial enquiry, and an account of the collective consumption process in a society like Britain's. In general, the examples come from the UK and the USA, although North American readers will find an imbalance to their disadvantage. Although this reflects my background and experience (rather than my prejudices), it also, of course, reflects the very different nature of the two societies, and particularly the primacy of state intervention in the British case. I have called the work *The Politics of Location* because I am keen to show that consumption issues have more than a static importance within social affairs (i.e. in terms of social status: Chapter 2). Throughout I have attempted to pick out types of social struggle and local conflict which reflect consumption questions. As already stated, the account is an introduction, an attempt to marshal and document the evidence that appears to support my thesis; the detailed task of explanation, and the even more taxing question of the potential political importance of these 'new' cleavages, now present themselves as the most pressing tasks.

References

Eyles, J. (1981) 'Why geography cannot be Marxist: towards an understanding of lived experience', *Environment and Planning*, A13,1371–88.

Kirby, A. M. (1982a) 'The external relations of the local state', in Cox K. R. and Johnston R. J. (eds) *Conflict, Politics and the Urban Scene*, London, Longmans.

Kirby, A. M. (1982c) 'Education, institutions and the local state', in Flowerdew R. *Institutions and Geographical Patterns*, London, Croom Helm.

Pahl, R. (1970) *Whose City?*, Harmondsworth, Penguin.

Saunders, P. (1981) *Social Theory and the Urban Question*, London, Hutchinson.

Preface to the Facsimile Edition 2015

The Politics of Location grew out of a course I taught at the University of Reading between 1977 and 1982. Parts were linked to research done earlier in the 1970s, focused on access to public services and the concentrations of poverty in the 'inner cities'. Against a backdrop of rising unemployment in the UK, fascist and anti-fascist marches and widespread urban decay, the goal was to identify inequalities manifested in particular places, and remedial policies that could be implemented via urban and regional policies.

Reading the book now, it is also easy to see a second concern, namely for disciplinary tensions within geography. This was a period of rapid change: Harvey's *Explanation* came out as I went to college, while its antithesis, *Social Justice*, emerged as I graduated—a definition of 'interesting times'.

In a post-postmodern era, it may be hard to grasp the sway that Marxist analysis held over the social sciences in the 1970s. Pushing back against it was not a way to gain recognition. But in fact almost any kind of social analysis was also likely to put one at odds with the traditional hierarchy within geography. *Politics of Location* received, I remember, some negative reviews from those who complained that it did not build on accepted foundations within political geography. In contrast, it was also critically reviewed by orthodox Marxists, who were offended that the book deliberately ignored production and related class struggles, and focused solely on Weberian concepts such as consumption, social status, the state and managerialism. Mine was a very minor precursor to the withering attacks experienced by Manuel Castells as his *Grassroots* was reviewed in 1983.

In short, the book was not Marxist enough for those interested in 'Politics', and it was not geographical enough for those interested in 'Location': nor was it technical enough for spatial scientists interested in GIS and more sophisticated ways of measuring accessibility. Reading the book now, one can see the earnest efforts to produce some balance between geography and a simple form of public administration or planning, but the tension is clear.

In reality, I was frustrated by the problems of working in geography even before this volume was concluded. The obligation to find the spatial dimension to every social issue sat uneasily with the need to find robust explanations rooted in state, civil society or market. Between 1982 and 1993 I focused on the development of a theory of the local state, which was eventually published as *Power/Resistance*. This was, I believe, a much more mature analysis. Yet it was again out of step with the academy. The social sciences have rarely put much effort into serious analysis of the state apparatus. And by the time my 1993 book was published, geography and planning were mired deep in the communicative turn and identity politics. It is not coincidental that I have not worked in a geography department since that time.

So, that might leave this volume as a relic, some social science tree rings that give clues to conditions of a former era. Yet, curiously, I note an uptick of interest in the issues that are at the core of the book. *Politics of Location*, and related publications dealing with welfare economics, are being cited again, as basic issues of access and delivery are back in fashion (although the context in which services are offered—or more often withdrawn—has changed dramatically in many countries).

So what is the take away for anyone looking at this book today? The deepening inequalities in almost all societies underscore the continued importance of wealth and class. These are also increasingly articulated via ethnicity and gender, and as I have intimated, the reader will find little insight here on this. Yet as inequality increases, it is the role of the local/state to provide for residents: to maintain the fundamentals of an acceptable quality of life. This book is not a blueprint for that process, but it does indicate the importance of **context** in that process.

Were I to re-write this volume today, I would also expand its scope to include the natural world. The pressing challenge of this century is the role of context with regard to rising sea-levels, earthquakes, hurricanes and tornadoes. Population increase, much of it in marginal urban locations, has placed hundreds of millions of people at increased risk from such threats, and has reduced the quality of their lives. The basic issues of service delivery have been inverted to become matters of **adaptation**. In this manner, the politics of location are even more pressing than ever before.

Andrew Kirby, Phoenix

References
Kirby, A. (1993). *Power/resistance: Local politics and the chaotic state.* Bloomington, Indiana University Press.
Kirby, A. (2014). Adapting cities, adapting the curriculum. *Geography, 99,* 90–98.

Part I

Space

1 A perspective on space

By virtue of the stress given by geographers to the *spatial distribution* of phenomena in *vacuo*, largely abstracted from their wider social contexts, they have unwittingly moved into an exploratory cul-de-sac productive of little more than often erroneous, spatially grounded, causal inferences, which simultaneously divert attention away from underlying social causes. (Chris Hamnett: *Social Problems and the City* (D. Herbert and D. Smith, eds))

Hamnett's views represent one side of a central debate within contemporary geography which revolves about no less than the efficacy of studying anything from a spatial perspective. In the past geography has faced a good deal of academic indifference – and hostility – and the move towards the creation of a professional discipline has been a painful process, as Johnston shows (1979a, pp. 86–99). However, the present debate is all the more important (and some might say dangerous) because it is being conducted *internally*. A sizeable number of geographers have become disillusioned with the subject, and they feel, to paraphrase Hamnett's argument, that a distinctly geographical approach can tell them less and less about anything of importance. To understand this standpoint more fully, we may usefully consider the ways in which space can be considered.

Space as an idea

The philosophical bases of space have been developed through several centuries:

in Kant's time there were, briefly speaking, two opposed conceptions of the nature of space. There was the viewpoint of the Newtonians in which space was treated as a real entity, with an existence independent of both mind and matter. Space was a huge container in which atoms and planets swam like fish in a tank. The view of Leibniz, however, was that Newtonian space was logically paradoxical. Empty space, clearly a nothingness, was, by the container conception, also a somethingness. This contradiction led Leibniz to believe that space was an idea rather than a thing: that 'space' sprang from the mind when thought conceived a relationship between perceived objects, and had no more real and independent existence than the distance between two persons described as near or distant relations. In this case, space was entirely relative: and if the objects were removed, space disappeared. (Richards, 1974, p. 3)[1]

Despite their antiquity, the views of space held within geography have tended to follow these 'relativist versus absolute' stances (see Harvey, 1969, pp. 206–9). Sack, for example, identifies two particular schools of thought, one of which he terms the 'spatial separatists', the other the 'chorologists' (1980, p. 330). The former, as the name suggests, holds 'that the spatial questions are about a separate subject matter – *space*; and that this subject matter required a separate kind of law or explanation – *spatial laws and explanations*' (1980, p. 330). Chorology is to be seen as a counterpart to chronology: 'the production of specific places, areas or regions, parallels the production of specific times such as an era or epoch in history' (1980, p. 331).

Chorology, naturally enough, examines particular places (absolute space(s)), and can draw upon any method or body of knowledge to assist in that study. The relativist view of space is, however, a very different animal, simply because it does view space as an object of study in its own right: thus, as Sack observes, there must be 'spatial laws' developed in order to understand its operation.

It was on this basis that post-war geography staked its claim to academic individuality, with what Sack terms 'spatial separatism'. Johnston in fact typifies much of the geography undertaken in the 1960s as being 'a lower-level science of spatial relations, applying in empirical contexts the laws of higher-level, generally more abstract sciences' (1979a, p. 98). Unsurprisingly, this concentration (some have described it as a fetishism) upon space as a 'thing-in-itself' has attracted detailed criticism. Sack himself has opposed separatism on philosophical grounds: simply, he argues that a unique geography must rest upon some innate properties of space, and these do not exist.

A more strident, and indeed wide-ranging attack, has come from an entirely different quarter, namely those who concentrate upon social and political issues. Some of the initial statements in this vein have in fact come from outside the subject, but the general thread of argument has been picked up and amplified by geographers themselves. In essence, the critique rests upon a rejection of the assumption that space can exist as an independent artefact, and that human spatial relations operate in the same way as atomic or planetary bodies. Soja observes:

while such adjectives as 'social', 'political', 'economic' and even 'historical' generally suggest, unless otherwise specified, a link to human action and motivation, the term 'spatial' typically evokes the image of something physical and external to the social context and to social action, a part of the 'environment', a context *for* society – its container – rather than a structure created *by* society. (1980, p. 210)

The argument here revolves around the assumption that space can be examined as a virtual abstraction. Such a notion is dismissed by two urban sociologists, Manuel Castells and Henri Lefèbvre. The latter writes:

If space has an air of neutrality and indifference with regard to its contents and thus seems to be 'purely' formal, the epitome of rational abstraction, it is precisely because it has already been the focus of past processes whose traces are not always evident in the landscape. Space has been shaped and moulded from historical and natural elements, but this has been a political process. Space is political and ideological. It is a product literally filled with ideologies. (1978, p. 341)

The same point is expressed with even greater succinctness by Castells: 'one runs the very great risk of imagining space as a white page on which the actions of groups and institutions are inscribed, without encountering any other obstacle than the trace of past generations' (1977, quoted by Soja, 1980, p. 212).

From these observations two inferences can immediately be made. The first is that space can only be understood as part of the operation of society; and the second is that space is used by society as an absolute thing: in other words it is of interest not in a relative sense (the distance between objects), but in the Newtonian context of a number of aquaria, within which we are located. Each portion of space has some particular importance (or meaning), and is used for a specific purpose: hence Lefèbvre's contention that space is a political reality. In turn then, one further inference can be made: if space is a social and political phenomenon, then there may be little point in studying it in the manner that we know as geographical; it might only be understood by those who concentrate solely upon social structures.

Spatial schema, spatial structure and spatial patterns

Although the argument proposed by Castells and Lefèbvre does undermine the idea of spatial separatism, it need not, however, be taken to imply that spatial considerations are always valueless.

Indeed, a good deal of the research undertaken within the broad realm of political economy has begun to explicitly examine the role of space in the functioning of the capitalist economy.

A particularly good example (albeit research within geography) is recent work done by Taylor, who provides a particularly stimulating analysis of how political geography can be approached using a political-economic perspective derived from Wallerstein (1980). Taylor's argument is essentially that, although the economic system is global in character, we must at times change the spatial scale when attempting to understand the operation of that world economy. His argument is illustrated in Figure 1.1.

This simple diagram suggests that although economic activity is global in character (the 'reality'), there exist in addition two subsidiary layers of understanding. The first of these is the nation state, which functions as an ideological[3] entity, a means of harnessing loyalties: 'nationalism is a mixture of idealist populism with hard-headed economic protectionism' (Taylor, 1980, p. 25). In contrast to the 'scale of ideology' is the urban, which is described as the 'scale of experience', 'the scale at which we live our daily lives' (Taylor, 1980, p. 25). It is at this level that, for example, the state provides consumption goods (housing and so on), and at this scale that the crises within capitalism are manifested; the redundancies and the closures. All three interact as follows:

In political discussion in the Wallsend Constituency in North East England I have observed that a major topic of concern has been the health of ship-building. This is only natural since the Swan Hunter yards are the major employer in the area. If the yards close, the resulting unemployment will affect the whole town, making Wallsend the 'Jarrow of the Eighties'. This is the scale of experience. It is at the scale of ideology that policy emerges however. The response to local pressures was for the Labour Government to nationalise British shipbuilding including Swan Hunter. This is ideological since it reflects only a partial view of the situation. It may protect jobs and ease the flow of state subsidies into the area, but it does not tackle the basic problem affecting shipbuilding. Both demand and supply in the Industry are global. The current problems in the Industry can be directly traced to the fall in demand following the 1973/74 oil price rise and the emergence of competitive suppliers from such countries as South Korea. Clearly a policy of nationalisation is a long way away from solving the problems of Wallsend's shipyards. (Taylor, 1980, p. 20)

Taylor's reading of Wallerstein provides us then with a rationale for examining phenomena at (at least) three spatial scales; not because it is an easy way to taxonomise or split up events, but because different things happen at different scales, and because it is difficult to examine something such as ideology at either the global

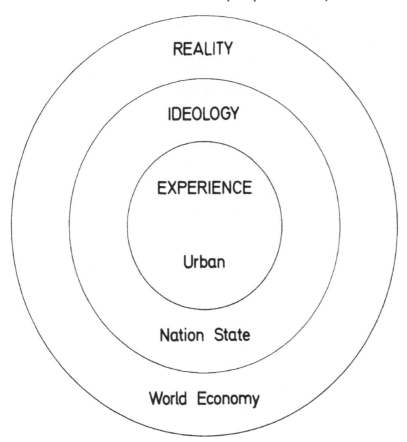

Figure 1.1   A rationale for the use of different spatial scales (after Taylor, 1980)

Although many geographical studies employ different scales of spatial resolution (international; national-regional; inter-urban; intra-urban), there is rarely any rationale given for these taxonomies. The emphasis upon nations and regions as obvious units of analysis is particularly worrying, as they are entirely artificial. Taylor's development of Wallerstein's ideas is one of the few attempts to critically account for the use of the particular scales 'national' and 'urban'; it is to be noted that the regional scale – described by Sayer as a 'chaotic conception' – is *not* employed.

or the local level. We can, however, take this a step further, in order to examine exactly why some activities are scale-specific, and in turn to test the effects of organizing any activity in a spatial domain. David Harvey, for example, has examined Marx's view of economic development, and he concludes that 'Marx recognised that capital accumulation took place in a geographical context and that it in turn created specific kinds of geographical structures' (1978, p. 263). Indeed, Harvey is of the opinion that Marx's outlines can be used to link up the process of capital formation with the emergence of imperialism, i.e. the diffusion of the capitalist system from one country to another: 'the theory of accumulation relates to an understanding of spatial structure . . . and the particular form of locational analysis which Marx creates provides the missing link between the theory of accumulation and the theory of imperialism' (1978, p. 263)

Harvey goes on to argue that distant lands serve to provide the core (of, for example, a nation) with three things: a surplus of labour, a supply of surplus capital, and a market for commodities produced in the core. Viewed in these terms, we can see that this argument not only implies the existence of geographically distinct economic activities – the spatial economy – but also that the spatial economy operates at different spatial scales:

in essence, depressed regions are presumed to be dependent colonies providing reserves of labour, capital and markets for exploitation by the core economy. The industrial reserve army (labour surplus) and the process of reproduction (of that labour) has thus a specific spatial character which inevitably involves uneven regional development. (Clark, 1980, pp. 226–7)

Similarly, Peet attempts to explain the existence of the inner city as a home for the reserve army (1975). It would be wrong, at present, to suggest that this 'spatial logic', or as Clark terms it, the 'necessity of (regional) inequality' argument, is not challenged; Clark's own paper, for example, represents such a critical review. None the less, it is clear that in this field the value of thinking in a spatial dimension is being vigorously examined.

It is, however, not enough to conceptualize 'spatial' and 'social' as phenomena or perspectives that flash in and out of focus, now together, now apart. As Gregory has demonstrated, there are inextricable links between 'spatial patterns' and 'spatial structures'.[4] He states:

It is clearly important to transcend geography's 'fetishism of areas' and 'to destroy the myth that areas, *qua* areas, can interact' (Carney, *et al.*, 1976), but it should now be equally obvious that this must mean more than a simple demonstration that the spatial lattice exhibits, in frozen and displaced form, a bundle of social relations . . . . The real problem . . . turns on the need to recognize (a) that spatial structures cannot be *theorized* without social structures, *and vice versa*, and (b) that social structures cannot be *practised* without spatial structures, *and vice versa*. (1978, pp. 120–1: original emphasis)

Gregory is thus suggesting that 'spatial' and 'social' are inseparable, and he has illustrated this by recourse to his now-familiar 'camshaft' analogy, which demonstrates that

spatial patterns cannot be theorized without reference to spatial structures, but also spatial structures cannot be operated without spatial patterns on which they are enacted. Thus, for example, whilst the internal spatial pattern of city residential areas reflects the class system in the spatial structure, that system is itself composed, in part, as a consequence of spatial patterns. (Johnston, 1979a, p. 164)

It is with Gregory's argument that we can most usefully confront Hamnett's views on geography's cul-de-sac, i.e. a spatial perspective. It is not enough to suggest that it is dangerous to divorce patterns from their 'socio-economic contexts'; it is in fact only when it can be logically demonstrated that pattern and structure (or more simply, space and society) are linked, that there can be any efficacy (as opposed to sterile academic interest) in geographical analysis at all. It is thus on this basic premise of Gregory's arguments that this book exists.

'People poverty' and 'place poverty'

The examples used above indicate that there can exist a spatial logic to the economic structures within society, be they regional economies or urban housing markets; more simply, spatial inequalities are actually part of the functioning of, say, a housing market, for without them, maximum returns cannot be extracted from particular parts of that market.[5] From this basic statement we can develop in turn a simple contention, namely that economic activities are likely to be reflected in space by social differences, be they variations in income between regions, or within urban areas: 'marked spatial differentiation of *per capita* income is therefore likely to be a permanent feature of the capitalist city' (Smith, 1977, p. 112).

This patterning is termed by Smith 'people poverty'; 'low-income people may occupy certain parts of a city by virtue of their low income, but their money incomes are not low because of where they live' (1977, p. 112). This he differentiates from 'place poverty', which 'emerges when other benefits or penalties compound the advantages or disadvantages of particular groups by virtue of where they live' (1977, p. 112). This implies a *direct* relationship between space and the individual, which we have as yet not considered.

The mechanisms by which place can influence the individual are varied, and analysis is not eased by the fact that some locations may appear to different observers in entirely different ways; (Leeds, for example, has been described by two contemporaries as 'a beastly place' and 'one of the grandest poems offered to the world';

(Pocock and Hudson, 1978, p. 1)). This problem notwithstanding, we may systematically attempt to build up a picture of 'place poverty'.

The aspect of the relationship between the individual and his or her environment which is easiest to understand is the transmission of *disbenefits*. Figure 1.2, which is a reproduction of a letter from a homeowner, provides a simple example. In it she complains of the behaviour of the children in a neighbouring block of flats; the list of complaints is self-explanatory, and reflects a clash of ages and lifestyles that is magnified by proximity.

These kinds of effects can be studied quite precisely, using the

---

Sunday 2 Sept. 9 p.m. Child screams fit to burst – one wonders, is he being murdered?
Monday Motorbikes revving long time, children noisy.
Tuesday Children noisy till dark.
Wednesday 5 Sept., 3.45. Ball game, very noisy, against some wall. 4.30 ditto, then mother yells at child. Child howls volubly.
Friday (Away at Bournemouth.)
Saturday 8th Sept. Child using noisy truck – excruciating. Mother shouts 'Philip' from top floor and generally creates noisy injunctions to the child. No peace . . .
Sunday 9 Sept. 4 p.m. Overloud horrible Indian music again from tenant Mr. Francis. He has been told about this half a dozen times, been written too . . . all to no avail.
Monday a.m. Mother shouting to child from upstairs. p.m. Football being played.
Tuesday 11 Sept. Mother yelling to kids. One mother shouts 'Simone' at frequent intervals. It is totally excruciating. No peace.
Saturday 15 Sept. Television on very loud. I can hear every word in my garden, three gardens away. It is totally unacceptable and must be stopped. I go to Mr. Francis and politely ask him to turn down his t.v. Later his wife sees us talking in the street and is thoroughly abusive. She shouts and it is all very upsetting. Her husband joins her. She is also rude to Mrs. Ward, whose husband so kindly took their child to the doctor, leaving his dinner, only a few days previously. I do think this sort of behaviour is disgraceful. We report it to Mr. Patton. We are now determined to seek a different type of tenant is installed . . .
Sunday 16 Sept continued. The dustbins smell to high heaven. They are also unsightly. We seek to have them screened off.
Poor weather, so there is a break.
Saturday 22 Sept. 10–11.30 a.m. Kids shouting, climbing all over neighbours' walls and the derelict building adjacent.
Tuesday 9 Oct. 6.45 p.m. Very noisy children. Also new tenant's boy sitting on top of washing line post, shouting. (As usual Mrs. Ward reports that children sat on their wall last night and disturbed them, staring in while they enjoyed their evening meal.)
This sort of incident demonstrates the strain the neighbours are suffering due to the presence of unsupervised children. Later there was no end of noise from children sitting on washing line post and yelling, also climbing on neighbours' sheds.

Figure 1.2 Letter of complaint from a homeowner to a landlord. The letter reveals a good deal of latent prejudice: concerning race, lifestyles and class. This notwithstanding, the complaints relate directly to the ways in which the neighbours' lifestyles affect the complainant due to their enforced proximity (she due to her age and 'reduced' economic circumstances, they due to the constraints of the private rental market).

concept of an *externality field*; this represents the spatial extent of the nuisance, and can easily be measured (Cox, 1973; Harrop, 1973; Smith, 1977). In Figure 1.3 the source of the externality is a football ground which periodically attracts large crowds, some of whom are noisy and occasionally violent. Large numbers of cars congest the nearby roads, hot-dog sellers pollute the air, and quantities of litter remain even when the match is over. The two maps of Figure 1.3 show the way in which these nuisances can be defined in terms of residents' perceptions (Bale, 1980, p. 94).

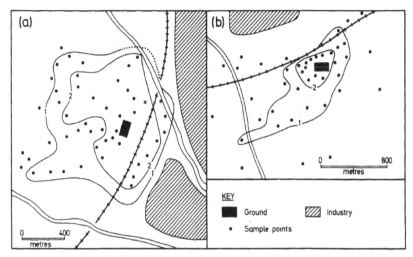

Figure 1.3   The measurement of the spatial extent of two externality fields; around (a) Derby County and (b) Charlton Athletic football grounds

Bale's investigation required respondents to rate the football ground in terms of a trichotomous scale: 0 = no nuisance, 1 = nuisance, 2 = severe nuisance. Similar distance–decay patterns have been observed with respect to externalities as dissimilar as rubbish tips and airports.

Clearly then, some locations may carry with them disbenefits. However, the football ground example reminds us that disbenefits have also to be balanced against benefits, and that apparently negative externalities may constitute positive externalities to some people. To football supporters, proximity to a ground may be beneficial, and crowds are desirable for shopkeepers and publicans. On a less tangible level football clubs also contribute quite markedly (through the rates) to the local revenue base, and they thus subsidize residents to some degree.

We can invert this argument, by concentrating in turn upon externalities which have a far more important positive element. In this context, we can identify some aspects of public provision, such as for example street-lighting, the quality of local roads and the maintenance of public open space. Although parks may attract noise and litter, they none the less improve an environment, both visually and practically, and in this instance, of course, the individual is not a passive actor, as (s)he is in the case of a football ground, where the nuisance is fixed and spreads only a certain distance. Where positive externalities are concerned, consumers can increase their benefits by travelling more widely, and taking in more facil-

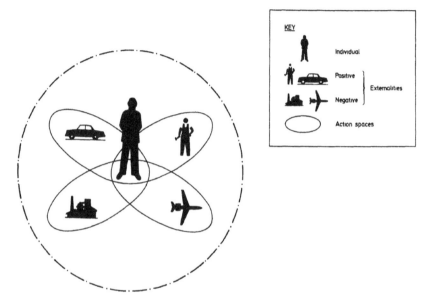

Figure 1.4   Individual action spaces and positive and negative externalities
The diagram is intended to represent stylistically several action spaces within which the central individual operates. Within each space exists an externality. Two of these are beneficial: a workplace and a motorway access point. Two are deleterious: a large chemical factory and a noisy airport. The overall quality of life is partly determined by the sum total of such externality fields; (note that for others, even these same sites may constitute different externalities: a motorway can, with proximity, lose its benefits (see Chapter 6), whilst an airport can, with some remoteness, represent simply an important transport mode (Chapter 5).

ities, such as, say, a museum or art gallery (and, of course, private aspects of provision, such as cinemas).

At this descriptive level therefore, we can view 'place poverty' in the way described in Figure 1.4. Here we see an individual located within a geographical space which contains two negative externalities and two positive ones. Also shown is our subject's *action spaces*, which simply reflect his or her preparedness to travel to various locations for leisure, for study, for employment or for social functions; clearly the more desirable the activity, the greater the

willingness to travel, and the action space will bulge accordingly.
Now the ability to travel will vary (with age and personal circumstances, particularly personal mobility and perhaps even the quality
of the roads or street lamps), but all things being equal, we can see
that what is available within the action space will do much to
determine the quality of life there. If the public and private facilities
are good, then these will, naturally enough, benefit the resident,
and perhaps outweigh any negative externalities. If the provision is
poor (and/or if the action space is very constrained) then disbenefits
may easily outweigh the few positive externalities: this then is 'place
poverty'.

The location of externalities

So far, we have considered the suggestion that space has a role to
play in the functioning of society, and that social differentiation may
be manifested as 'people poverty', which varies from place to place.
This we have contrasted with 'place poverty', whereby the individual household is affected by the externalities about it.

What we must not now assume is that the direct impact of location
upon the individual is a simplistic issue, somehow different from
other processes at work within society. We cannot take for granted,
for example, that externalities are regularly spaced upon the landscape (i.e. that noxious sites, such as power stations, are placed to
serve every few hundred square miles of territory), and that disbenefits are thus evenly distributed. Nor can we assume that
benefits are randomly distributed throughout, say, a city: that, in
other words, poor neighbourhoods are as likely to possess a
recreation site as is a rich one, or that black communities are as
likely to have access to a library as is a white community.

Such assumptions are untenable on two counts. The first is that
the distribution of population is not a fixed and constant thing;
rather, it is a constantly shifting set of patterns, with large numbers
leaving areas such as inner cities in preference for suburban and
even non-urban areas. The reasons why such movements have
occurred are numerous, but in part at least they relate to the
superior benefits available in suburban areas:

housing units in inner city areas are often the least attractive units in the
least attractive neighbourhoods; they also tend to be in jurisdictions where
tax-rates are relatively high and where education is short-shrifted, so that
more money can be spent on public safety. Housing units in suburban areas
are usually much newer . . . and located in less densely occupied neighbour-

hoods, with good schools, few public safety problems, and relatively low tax rates. (Cox, 1979, p. 276)

Using this kind of information, we can therefore argue that some households (the most affluent) can relocate themselves at will in order to avoid 'place poverty' (Tiebout, 1956).

The second factor to take into account is on the other side of this coin, namely that not only do populations migrate, but externalities themselves are continually improved, enlarged and relocated. Falling rolls may cause schools to be closed in some areas, whilst redevelopment elsewhere attracts new facilities. Disbenefits (such as gas works) disappear due to technological changes from inner areas, to be replaced by nuclear power stations in rural areas." In each case, the externality in question may be the focus of local political concern, with different communities competing for the benefits, and attempting to avoid the acquisition of disbenefits. As Harvey has observed: 'much of what goes on in a city (particularly in the political arena) can be interpreted as an attempt to organise the distribution of externality effects to gain income advantages' (Dear, 1979, pp. 56–7; Harvey, 1973, p. 58; Kirby, 1979, pp. 346–8).

Taken together, these two factors (population mobility, and the creation and subsequent demise of externalities) imply that 'place poverty' is not a random occurrence. This inference is strengthened if it is realized in addition that many aspects of public provision (i.e. the benefits) are not distributed via some aspatial strategy (e.g. one doctor per 2000 people), but via a network of geographic units. These *legally bounded spaces* exist at many spatial scales and are under the jurisdiction of very different authorities, as Table 1.1 indicates (see also Coates, Johnston and Knox, 1977, pp. 182–225).

The importance of the juridical context, as Cox terms it, lies in the fact that it adds an additional dimension to 'place poverty'. Thus, rather than considering simply the presence/absence of externalities and the ways in which these impact upon an individual, we can also compare the *quality* of a particular benefit and the way in which the level of provision varies from one spatial unit to the next (this was the point made by Cox above with respect to the quality of schools in suburbs and inner cities). In this sense, then, we are moving away from a simple model which considers externality fields, towards a rather more sophisticated approach which takes into account the level of services provided within a legally bounded space, and even the costs borne by individual households in pay-ment for their services. As we shall see in Chapter 3, levels of provision do vary enormously in fields like health and education, whilst the costs imposed upon populations also differ, but not necessarily proportionately. Both Cox and Johnston document the ways in which the jig-sawing of the local government map can lead to concentrations of low-income households, with pressing require-

Table 1.1 Administration dependent upon spatial units, England and Wales

Administrative unit (and number)	Function
Regional health authorities (15)	Expenditure upon health care
Water authorities (10)	
Gas boards (12)	Provision of utilities
Electricity boards (12)	
Rent assessment panels (15)	Determination of Fair Rents for tenants
Police constabularies (43)	Organization and provision of police services
Greater London Council	Highway and housing provision in London
Counties (53)	Highways
Metropolitan counties (6)	
Districts (333)	Housing and social services
Metropolitan districts (36)	
London boroughs (32)	Housing, transport
Local valuation panels (67)	Dealing with rating and property issues
Local education authorities (104) (excluding London)	Expenditure and organization of school systems
Area health authorities (99) (now replaced by district health authorities)	Expenditure on health care, particularly primary care
Wards } total 16,573	Electoral division, sub-urban scale
Parishes }	Electoral division, rural areas
There exist in addition various ad hoc units	
Postal districts	Mail delivery
Police divisions	Organization of local law enforcement
High Sheriffs' shrievalties	Distribution of writs

ments for services, in bounded areas where the local revenue base is low, and this is in general a function of the old age of the property involved. Suburban residents, conversely, are able not only to raise adequate funds for their own requirements, but are frequently able to make use of 'downtown' facilities in neighbouring cities (Cox, 1979, pp. 255–69; Johnston, 1979b, pp. 37–44)

The politics of collective consumption

Clearly then, the provision of goods, the location of disbenefits and the manipulation of local government (and other) boundaries are all politically 'loaded' issues, with important consequences both for areas and the people who live within them. Indeed, recent developments in urban analysis point to the conflicts that develop with respect to provision as being ultimately of great, and occasionally of even greater importance, than conflicts which develop within the workplace; simply, issues dependent upon consumption (of housing, education, health care) are potentially as important as the class issues that are dictated in the area of production, as these cleavages become expressed as political support for different electoral parties:

[a]   consideration of the economic, ideological and political structures created by state intervention in housing and transport generates two fairly clear conclusions about the patterns of consumption cleavage in present-day Britain. First, there are consumption cleavages which cannot be assimilated into the dichotomy between 'middle class' and 'working class' and which are distinct from both production cleavages and from pre-dominantly ideological issues. Secondly, patterns of housing cleavage revolve not around the ownership or non-ownership of domestic property, but around collective versus individualized consumption. Nor is this housing cleavage unique, for there are clear parallels with that of transport. (Dunleavy, 1979, p. 436)

Let me explain. Dunleavy is arguing here that certain groups (such as homeowners and car owners) are several times more likely to vote Conservative than are their counterparts who live in council homes or who are without private transport, *even when social class is held constant*. The conflict is, moreover, not one between home-owners versus non-homeowners, but specifically between council tenants (i.e. those in public housing) and homeowners, for relative increases in their respective subsidies: homeowners receive mortgage relief, whilst council tenants receive rent subsidies and rent rebates.

As he has outlined his argument, Dunleavy is simply monitoring a shift in national electoral cleavages, away from ideological bases and towards 'new' issues such as consumption. This is, however, to take a limited view of the political process. Castells, for example, has documented at great length the ways in which consumption issues can generate spontaneous outbursts of popular indignation within neighbourhoods where publicly provided benefits (or 'the collective means of consumption') are threatened by either public or private actions:

Any investigation in a neighbourhood or suburb will reveal a great number of small, day-to-day struggles involving [the] means of collective consumption. Associations are much richer than one would think, and residents are quite sensitive to problems of the quality of life. In any case, they are as sensitive to these problems as they are to political and economic questions in general. (1978, p. 148)

Castells's arguments are of interest to us in several ways.[1] Of particular importance are the suggestions that, first, 'social cleavages [may be] related to the accessibility and use of certain collective services, from housing conditions to working hours, passing through the type and level of health, educational and cultural facilities' (1978, pp. 15–16). Second, these cleavages are socially widespread, in that

a crisis in the provision of the means of collective consumption will hit the working class and middle class alike; the middle class need hospitals, they use public transport such as rail commuter links, they send their children to state schools (or, at least, some of them do). It follows that new patterns of inequality arise in the urban system which do not correspond with those generated in the work situation. (Saunders, 1980a, pp. 112–13)

Third, these cleavages are frequently a function of a failure of provision in a specific place: 'they are locally based, in which case they tend to be locality-specific, issue-specific and strategically limited' (Saunders, 1980b, p. 50).

Castells himself has argued that these more limited conflicts may constitute a foundation for what he describes as 'urban social movements', which may in turn represent an alternative to existing political aspirations, which he dismisses as the 'political ghetto' or the 'ideological utopia'; in other words, they may form the basis for revolutionary struggles. The extent to which this chain of argument is feasible is open to debate; certainly, it appears to be the case that there have been no examples of dynamic urban social movements within advanced societies in recent years, and both Saunders and Pahl are very sceptical of such logic being usefully applied to the British situation at all (Pahl, 1978, p. 314; Saunders, 1980a, p. 17). This is, however, not to undermine the efficacy of Castells's general point, which we can illustrate with empirical evidence.

In their investigation of locational conflicts in Canada, for example, Ley and Mercer observe that

land-use controversies during the 1970s have focused upon questions like freeway construction, redevelopment, and the provision of public goods such as open space – questions which have frequently been resolved through political initiative or through the courts. In this manner, locational decision-making (and locational conflict) is increasingly being mediated through *political processes* and not simply the operation of the market. (Ley and Mercer, 1980, p. 91, original emphasis)

Using data amassed from the newspaper *Vancouver Sun*, the researchers have documented the relative importance of various issues, and the protagonists involved (Table 1.2).

As we can see, several clear patterns emerge. First, issues such as transportation, housing and redevelopment appear as far more

Table 1.2 Major characteristics of locational conflicts

A Conflict type	No. of issues	**B** Proposed land use	Per cent[a]
Transportation	27	Transportation	32
Recreation	15	Housing	32
Redevelopment	14	Commerce	28
Commerce	12	Parks/open space	19
Housing	12	Public services	14
Public institution	7	Industry	3
Preservation	5	Unknown	2
Noxious facility	4		
School	2		
Total	98		

C Participants	Initiators (per cent of issues)[a]	Advocates (per cent of issues)[a]	Opponents (per cent of issues)[a]
Households	11	33	54
Neighbourhood groups	6	19	31
Civic organizations	3	15	32
Land developers	28	32	5
Other entrepreneurs	5	9	10
Elected officials	26	56	48
Administrative officials	18	31	33
Unknown	3	0	0

D Conflict intensity	Per cent of issues[a]
Verbal disagreement (face-to-face)	77
Verbal disagreement (indirect)	72
Appeal to government	53
Group organizing	48
Petition	29
Brief presented	21
Demonstration	20
Injunction served	6
Arrests	0

Table 1.2 Major characteristics of locational conflicts – *continued*

E *Grounds for involvement*	*Advocates* (per cent of issues)[a]	*Opponents*
(I) *Economic factors*	13	19
Cost of change	5	7
Impact on property values	1	6
Other economic factors	7	6
(II) *Social factors*	38	67
Need for improved services	11	9
Compatibility with neighbourhood	2	12
Impact on traffic	3	12
Safety	15	3
Availability of housing	3	9
Other social factors	4	22
(III) *Aesthetic factors*	14	22
Visual attractiveness	8	14
Other aesthetic factors	6	8
(IV) *Procedural factors*	6	15
(V) *Miscellaneous factors*	2	13

F *The geographical involvement of conflict groups*

	CBD Conflicts Pro Con		*Inner city* Westside Conflicts Pro Con		Eastside Conflicts Pro Con		*Inner suburbs* Westside Conflicts Pro Con		Eastside Conflicts Pro Con	
Households	4	9	11	13	1	2	7	18	9	8
Neighbourhood groups	2	3	8	11	0	3	4	9	3	3
Civic organizations	4	9	3	11	1	1	5	7	1	3
Land developers	10	1	7	3	4	1	6	0	4	0
Other entrepreneurs	3	5	4	6	1	0	2	1	0	2
Elected officials	14	17	12	9	5	3	15	12	8	5
Administrative officials	5	6	12	7	1	4	7	7	3	8

Source: adapted from Ley and Mercer (1980).
[a] Multiple counting prevents per cent from summing to 100.

conflict-ridden issues than, say, the distribution of noxious industries. Secondly, conflicts are typically, although by no means exclusively, expressed as residents versus politicians and officials. Thirdly, a sizeable proportion of issues take on some importance, involving petitioning and even demonstrations; and fourthly, types of conflicts are likely to be geographically concentrated, with organizations being more common in the inner city, and individual households more common as opponents in the suburbs. (These tendencies, and other issues, are explored further in Part III.)

To conclude this section we may usefully consider the directions in which this argument is taking us. We have now proceeded far beyond a simple consideration of externalities, towards the suggestion that consumption struggles may, in certain localities, come to mould the political process, and to augment and even oust more traditional cleavages. Clearly, an emphasis solely upon externalities is far too slim a base upon which to build such an important argument, and it is necessary that we develop a broader method of understanding the relationship between the spatial organization of society and its impacts upon the individual. This is a task attempted in the following chapter.

Conclusion

In this chapter I have attempted to do four things: first, to generally assert that it is important to examine society in a spatial domain; secondly, to consider the way in which certain activities (such as the housing market) take on the forms that we recognize, *inter alia* by virtue of their geographical locations. Thirdly I argue that, in addition, location may determine certain aspects of the individual's quality of life, as a result of the externalities that exist there; and fourthly, and lastly, I try to show that locationally based issues are widely seen as being of importance, and may assume a political importance as a result. All of these questions are amplified below.

Notes

1 Kant, himself an amateur geographer, viewed space as 'a schema for co-ordinating with each other absolutely all things externally sensed', a geographical 'stage on which we shall present all experience' (Richards, 1974, p. 7).
2 The use of the term 'ideology' here is a complicated one. A simple alternative would be 'meaning', implying that all space has meaning with respect to some form of land use.
3 The reader will have noted that this use of the term 'ideology' differs from that outlined by the author in note 2 above. The problem is simply stated, in so far as there exist two distinct uses of the word. The first relates to its *positive* meaning, in the sense that an ideology is associated with a particular class or group, and thus constitutes a 'world-view'. The second relates to its *negative* meaning, in the sense that an ideology constitutes a delusion or false-consciousness. Here, Taylor is using the second meaning, to suggest that nations themselves are responsible for per-petuating the assumption that nation states are the natural order of geographical organization (1980, p. 21). For a simple introduction to ideology, see Kirby (1980); for a detailed analysis, see Larrain (1979).

4 There exist, too, relations with 'spatial schema', which can be interpreted as basic neural responses to the world (Gregory, 1978, pp. 99–104).

5 For example, the most expensive housing in an area must necessarily be spatially concentrated, in order to maintain exclusivity. A market based solely upon the age and quality of dwellings would be very different from one in which certain types of property are grouped into localities, allowing some economic and/or ethnic groups to maintain a 'strategic distance' from their counterparts.

6 In this example the movement of disbenefits is away from the inner area; this, however, also implies a shift in employment opportunities (Community Development Project, 1977).

7 I have thus far avoided discussion of one of the more contentious aspects of Castells's work, namely his emphasis upon the so-called 'specificity of the urban', whereby he examines the existence of urban areas and their role within capitalist society. He argues that cities function primarily as stores of consumption goods (hospitals, schools, houses) that enable the labour force to successfully survive. This not only implies that issues of social organization have a spatial expression, but that urban areas are, in various ways, different from non-urban areas: that there are, for example, specifically 'urban problems', that develop when the provision of goods like housing fails. For a particularly good critique of this specificity assumption, based upon the argument that cities are not simply designed to meet consumption needs, see Saunders (1980b).

References

Bale, J. (1980) 'Football clubs as neighbours', *Town and Country Planning*, 49(3), 93–4.

Carney, J., Hudson, R., Ive, G. and Lewis, J. (1976) 'Regional under-development in late capitalism', in Masser, I. (ed.) *Theory and Practice in Regional Science*, London, Pion.

Castells, M. (1978) *City, Class and Power*, London, Macmillan.

Clark, G. (1980) 'Capitalism and regional inequality', Association of American Geographers, *Annals*, 70, 226–37.

Coates, B., Johnston, R. J. and Knox, P. L. (1977) *Geography and Inequality*, Oxford, University Press.

Community Development Project (1977) *The Costs of Industrial Change*, London, CDP.

Cox, K. R. (1973) *Conflict, Power and Politics in the City*, New York, McGraw-Hill.

Cox, K. R. (1979) *Location and Public Problems*, Oxford, Blackwell.

Dear, M. J. (1979) 'Thirteen axioms of a geography of the public sector', in Gale, S. and Olsson, G. (eds) *Philosophy in Geography*, Dordrecht, Holland, Reidel.

Dunleavy, P. (1979) 'The urban basis of political alignment: social class, domestic property ownership and state intervention in consumption processes', *British Journal of Political Science*, 9, 409–43.

Gregory, D. (1978) *Ideology, Science and Human Geography*, London, Hutchinson.

22   The politics of location

Harrop, K. J. (1973) 'Nuisances and their externality fields', *Seminar Paper 23*, Department of Geography, University of Newcastle-upon-Tyne.
Harvey, D. W. (1969) *Explanation in Geography*, London, Arnold.
Harvey, D. W. (1978) 'The geography of accumulation', in Peet, R. (ed.) *Radical Geography*, London, Methuen.
Johnston, R. J. (1979a) *Geography and Geographers*, London, Arnold.
Johnston, R. J. (1979b) *Political, Electoral and Spatial Systems*, Oxford, Oxford University Press.
Kirby, A. M. (1979) 'Public resource allocation: spatial inputs and social outcomes', in Goodall, B. and Kirby, A. M. (eds) *Resources and Planning*, Oxford, Pergamon.
Kirby, A. M. (1980) 'An approach to ideology', *Journal of Geography in Higher Education*, 4(2), 16–26.
Larrain, J. (1979) *The Concept of Ideology*, London, Hutchinson.
Lefèbvre, H. (1978) 'Reflections on the politics of space', in Peet, R. (ed.) *Radical Geography*, London, Methuen.
Ley, D. and Mercer, J. (1980) 'Locational conflict and the politics of consumption', *Economic Geography*, 36(2), 89–109.
Pahl, R. (1978) 'Castells and collective consumption', *Sociology*, 12(2), 309–15.
Peet, R. (1975) 'Poverty and inequality: a marxist-geographic theory', Association of American Geographers, *Annals*, 65(4), 564–71.
Pocock, D. and Hudson, R. (1978) *Images of the Urban Environment*, London, Macmillan.
Richards, P. (1974) 'Kant's geography and mental maps', Institute of British Geographers, *Transactions*, 61, 1–16.
Sack, R. (1980) 'Conceptions of geographic space', *Progress in Human Geography*, 4(3), 313–45.
Saunders, P. (1980a) *Urban Politics*, Harmondsworth, Penguin.
Saunders, P. (1980b) 'Towards a non-spatial urban sociology', *Working Paper 21*, Urban and Regional Studies, University of Sussex.
Smith, D. M. (1977) *Human Geography, a Welfare Approach*, London, Arnold.
Soja, E. W. (1980) 'The socio-spatial dialectic', Association of American Geographers, *Annals*, 70(2), 207–25.
Taylor, P. J. (1980) 'A materialist framework for political geography', *Seminar Paper 37*, Department of Geography, University of Newcastle upon Tyne.
Tiebout, C. (1956) 'A pure theory of local government expenditures', *Journal of Political Economy*, 64, 416–24.

2 Deprivation

I'd say deprivation was going on right now.
(Malcolm Bradbury: *Stepping Westward*)

Consumption and deprivation

In Chapter 1, under the heading 'The location of externalities', we touched upon some of the control issues that lie behind the simple spatial patterns that are observed, and examined briefly the political issues that govern the distribution of benefits and disbenefits upon the landscape. If we follow up Castells's arguments further, however, we are placed in a position in which we need to answer certain additional questions. For example, why are certain benefits provided within the public arena at all?

Castells's thesis is an interesting one because it lashes together a whole series of similarly fundamental questions, and provides a framework for their understanding. In simple terms, a direct link is made between the state, on the one hand, and a housing scheme or a new rapid-transit system on the other. The state, it is assumed, intervenes to provide these consumption goods because the market is unable – or unwilling – to do so, and without them social relations would collapse: children would have no schools, workers could not get to work, and families might not have homes.

This argument, which forms the basis of much debate in contemporary urban sociology, is important in several respects, and not the least of these is that it gets political analysis out of the groove 'in which it has long seemed stuck' (Miliband, 1973, p. 7). In particular, it broadens the study of the state away from an emphasis upon economic issues (production), and introduces, as we have seen, the other side of the coin (consumption). This is not to say, of course, that the thesis is without its critics. The definition of 'collective consumption', the specificity of the urban and the existence and form of urban social movements have all been attacked (see Dunleavy, 1980, pp. 42–50; Pahl, 1978; Saunders, 1980 pp. 113–27).

As far as the argument being developed here is concerned,

Castells's general thesis is taken as a useful account of the political and social importance of public provision, and the social conflicts that it engenders. It is, however, to a degree, only a partial argument. For example, as the logic is established in diagram (a), one important aspect is overlooked: the individual, and the individual's quality of life as it is affected by benefits and disbenefits, is not assessed. In so far as 'place poverty' is addressed at all, it is seen only as a catalyst for political action. Any additional links are not explored, although diagram (b) indicates how these might be viewed:

(a)

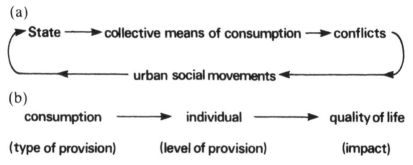

(b)

consumption   ———▶   individual   ———▶   quality of life

(type of provision)   (level of provision)   (impact)

Nor is this the only drawback: a further term is missing from the locational equation. We have as yet no explicit statement as to the factors that actually govern the individual's spatial actions, i.e. how (s)he comes to be in a particular location:

(c)

production   ———▶   individual   ———▶   location

(job)   (income)   (segregation)

In short, and to reintroduce the terms used in Chapter 1, we have only a limited account of 'place poverty', and no account at all of 'people poverty'. Nor is this simply an urge to create order where none necessarily exists. An assumption implicit within Castells's argument is that consumption is a process that is in some way separate from production; expressed another way, this suggests that whilst some people are involved in issues over transportation, others are being faced with redundancies and factory closures, and that in consequence workplace issues can be seen as totally distinct from geographical (externality) issues. But is this necessarily the case? Is it not likely that 'people poverty' is in some way connected to 'place poverty'? Are not, simply, the poor areas of cities also likely to be the ones to face urban renewal or the loss of health-care facilities? And if this is the case, why should we bother focusing upon 'place poverty', when a study of more traditional forms of inequality will allow us to identify those under threat far more quickly?

This chapter is an attempt to confront this question of the relationships between the effects of the operation of the market place (that we can, in shorthand, term social class), and the results of issues specifically to do with space and consumption. There are several ways to approach this, but here I want to examine the notion of *deprivation*, and to focus particularly therefore upon individuals and groups who are poorly placed within society; as it is here that the issues are revealed most sharply.

The concept of deprivation

Most discussions of 'deprivation' begin with some reference to the problems of defining such a concept (see, for example, Hirschfield, 1978; Rutter and Madge, 1976). The latter authors, using an allied term 'disadvantage', question whether one should use a statistical or an administrative definition, or turn instead to criteria that involve self-perception or individual values (Rutter and Madge, 1976, p. 8). As far as the perceptions are concerned, the work of Runciman has clearly shown that individual views of disadvantage (as related to, say, income) bear little relationship to the absolute level of deprivation (Runciman, 1966). Similarly, the problems of arriving at some objective, statistical definition of the concept are immense, as has been noted with respect to the inner-city issue of 'multiple deprivation' (see, for example, Holtermann, 1975; Townsend, 1979).[1]

A more satisfactory approach, in one way at least, is to eschew definition (the symptoms of deprivation) and to turn instead to the *mechanisms* by which disadvantage comes about. This takes for granted that deprivation exists, and given the focus of this argument, that it exists in a spatial context. Leaving aside for the moment the extent to which different types of deprivation overlap in space (Hamnett, 1979; Kirby, 1978), the material presented in later chapters clearly indicates that at a whole series of spatial scales, there exists wide variation along many axes of measurement such as income, housing quality and educational attainment.

The causes of deprivation

There exist several ready-made theories as to how deprivation arises, and this issue is rehearsed by Rutter and Madge in enormous detail, although they concentrate upon subject areas (income, crime) rather than the different models *per se*. The topic is discussed in less detail by Hirschfield, but his analysis is an explicit one of competing causes, and the range of explanations he presents is outlined in Table 2.1.

Table 2.1 A typology of social problems

Concept	Causal model	Explanation of the problem
Poverty	Culture of Poverty	Arising from the internal pathology of deviant groups
Deprivation	Cycle of Deprivation	Arising from individual psychological handicaps and inadequacies transmitted from one generation to the next
Disadvantage	Institutional malfunctioning	Arising from failures of planning, management or administration
Underprivilege	Maldistribution of resources and opportunities	Arising from the inequitable distribution of resources
Inequality	Structural class conflict	Arising from the divisions necessary to maintain an economic system based on private profit

Source: adapted from Community Development Project (1975); Hirschfield (1978)

Of the five concepts presented in the table, some have unique meanings: 'poverty', for example, possesses a distinct image: we talk, for example, of 'the poverty line', in the sense of a particular level of income. Some are, on the other hand, interchangeable: 'deprivation' and 'disadvantage' have already been so used above, although 'underprivilege' is virtually an oxymoron, and would be better rendered as 'not-privileged'. Because of these imprecisions it is easier to concentrate upon the three clear processes on display, which represent (in mathematical terminology) three disjoint sets, as outlined in Figure 2.1. It is these three types of *process* that will be examined here rather than the five concepts.

The first process relates to personal abilities, and more particularly inabilities, to cope with day-to-day existence. It has been related back to, on the one hand, the emergence of subcultures and the rejection of the 'normal' ethos of society (such as by Rastafarians), or on the other hand a transmission of deprivation from parents to their children. Rutter and Madge's work came into being as an attempt to gauge the efficacy of such views, which have had some political popularity, and their conclusion is that:

|an| apparent focus on the family is too narrow . . . continuities over time regarding high rates of various forms of disadvantage can be seen in terms of schools, inner city areas, social classes, ethnic groups and other social and cultural situations which lie outside the family. (1976, p. 6)

Naturally, a total rejection of individual factors is unwise; however, given the aggregate scale of this argument, no further

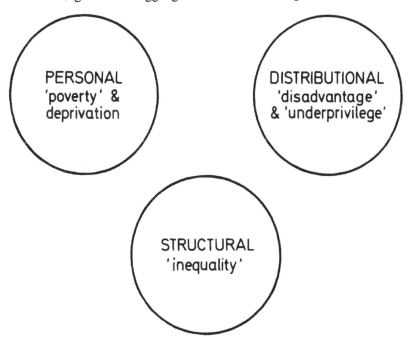

Figure 2.1   Three processes leading to deprivation: personal, distributional and structural. The terms in inverted commas relate to Table 2.1, and are used there to define the outcomes of the three processes

consideration will be given here to the personal bases of deprivation, and our use of the latter term does not imply such a perspective.

Structural inequalities

By focusing in turn upon structural explanations of disadvantage, we move from models that single out the individual to one that considers the entire economic and social system: Peet for example succinctly writes:

the Marxist view is that inequality is inherent in the capitalist mode of production. Inequality is inevitably produced during the normal operation of capitalist economies, and cannot be eradicated without fundamentally altering the mechanisms of capitalism. In addition it is functional to the system, which means that power holders have a vested interest in preserving social inequality. There is little point therefore in devoting political energies to the advocacy of policies which deal only with the symptoms of inequality without altering its basic generating forces. (1975, p. 564)

As Peet observes, Marxism (which is not the sole, but is the most coherent method of analysis in this context) is a 'meta-theory', in so far as it attempts to make sense of the globe; as such it is aspatial, and can be applied at urban, national and international scales. As he continues:

geographic theory deals with the mechanisms which perpetuate inequality from the point of view of the individual. It deals with the complex of forces, both stimuli and frictions, which immediately shape the course of a person's life. It is the perfect microscale complement to the macroscale Marxian analysis. (p. 569)

In terms of evaluating a particular model of deprivation, Peet is thus suggesting some marriage between a meta-theory (termed above 'structural') and geographic theory (termed 'distributional'). The latter nests with the former as follows:

the individual's struggle to earn income takes place in a certain physical, social, and economic environment. This environment may be thought of as a set of resources – services, contacts, and opportunities – with which the individual interacts. Central to the idea of geography of inequality is the realisation that a person may only exploit the social resources of a limited section of space in order to ready himself for the labour market. (p. 569)

Peet is therefore offering a sophisticated view of inequality (as the term is used in Table 2.1) that includes not only the tensions that result from class conflict in the economic domain, but the related issue of resource consumption. In short, 'people poverty' is equated *directly* with 'place poverty'.

It should be noted here straight away that Peet is not approaching the distributional issue in exactly the same way as it has been tackled above. For Peet, the consumption question is simply a subset of an all-embracing economic one; resources, such as education, are exploited in order that the labour market can be successfully – or unsuccessfully – entered. Consequently, a close circular relationship is seen to exist between economic and resource issues: 'each age-group, each social class, each racial group, each sex, has a different sized typical daily "prism" in which to operate. For the lowest class and most discriminated-against groups, the prism closes into a prison of space and resources' (Peet, 1975, p. 569).

This view is limited on two counts. First, it assumes a cyclical process, whereby the poor are unable to reinvest in the 'resource environment', and are thus trapped in the worst examples of that social environment. This, however, assumes that all the legally bounded spaces responsible for the creation and maintenance of that environment are isomorphic with rich or poor neighbourhoods, which is difficult to demonstrate (see later under the heading 'The mechanisms of deprivation'). Secondly, it also ignores the existence of benefits and disbenefits – recreational space and noise, for instance – which change the quality of life, but which do not link back directly into the labour-market argument.

These two problems suggest that the distributional question cannot simply be subsumed into the structural argument, as Peet proposes. The problem remains, however, as to how it is to be approached.

Distribution and social status

In Table 2.1, two concepts are proposed to account for the maldistribution of resources. The first, 'disadvantage', is a little unhelpful, in so far as the proposed mechanism – institutional malfunction – suggests an idiosyncratic process rather than an observable process. Clearly, such malfunctions occur, but it is impossible to build up any systematic account of such phenomena.² The alternative is 'underprivilege', which implies an inequitable distribution of resources.

As we have seen, one possibility is to tie these distributional issues directly into the structural one. An alternative does however exist.

Weber's sociology may be said to constitute an attack upon the Marxian generalisation that class struggles form the main dynamic process in the development of society. This theorem is questioned by Weber, in two main respects: first that by seeing the 'political' as secondary and derivative, it greatly exaggerates the significance of 'economic' relationships . . . second, that it fails to recognise the part played in history by *status affiliations, created as bases of group formation through processes which are not directly dependent upon class relationships.* (Giddens, 1973, p. 50; my emphasis)

Let me explain. The Weberian schema identifies two axes within society. The first, naturally enough, is economic class. The second is, crudely, a 'style of life'. It is however not simply a subjective issue, a matter of choice:

with some oversimplification, one might thus say that classes are stratified according to their relations to the production and acquisition of goods; whereas status groups are stratified according to the principles of their consumption of goods represented by special styles of life. (Weber, 1948, p. 180)

What we have here, then, is a second, distinct set of processes at work. Clearly, at times the axes will converge, at others they will not:

status affiliations cut across the relationships generated in the market [although] classes and status groups tend to be closely linked through property: possession of property is both a major determinant of class situation and also provides the basis for following a definite style of life. (Giddens, 1973, p. 44)

If we translate Giddens's remarks into a spatial dimension, we can see that, *on some occasions*, 'people poverty' will overlap with 'place poverty', but that this is not a conditional relationship: one does not depend upon the other. What is more likely is that different status groups will emerge, reflecting the quality of life available within different locations, *even when social class is held constant*. Some groups will exploit leisure facilities, some transport opportunities, others the educational provision, and in each context, the groups will possess a subtly different style, and quality of life.

To summarize this section, we are here confronting an argument that suggests that, on the one hand, we have individuals differentiated by structural factors, and that this will be reflected geographically, usually by the housing market. These social classes may then be further differentiated by distributional factors, which will also have a spatial manifestation; these will add to the process of creating status groups, each with a different quality of life.

The structural questions alluded to are relatively well documented, not least in the context of the housing market (Bassett and Short, 1980). What are less well understood are the detailed issues of resource distribution, and it is these that we will now attempt to systematically examine.

The measurement of distributional effects

First of all it is necessary to build up some coherent picture of resource provision. The time–space prism has been introduced, but for various reasons this is only of limited efficacy here. The concept of a prism is useful in so far as it underlines the spatial limits of accessibility; simply there is a finite distance that individuals can, or are prepared to, travel to a particular 'station' such as a school or a clinic (Thrift, 1977). Furthermore, it is useful to remind ourselves that accessibility is not necessarily the same as 'effective access' (Ambrose, 1977): in other words, the temporal constraint may be insurmountable. In terms of Peet's discussion, an example might be the limitations placed upon female labour by the temporal constraints of school hours, which dictate part-time, rather than full-time employment for many women.

Beyond this, the space–time approach cannot really go (in measuring the variation in the provision of resources between prisms for example), although work undertaken by David Smith suggests one way that this can be approached.

Smith has drawn deeply on welfare economics, which, suitably converted, provides a means of examining 'who gets what, where and how', in the sense outlined in Figure 2.2. This is a repre-

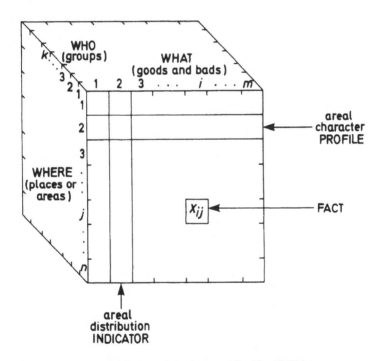

Figure 2.2   A welfare data matrix (source: Smith, 1977b)

The cube is an adaptation of Berry's geographical data matrix, and is designed to provide a simple descriptive account of welfare variations. It can be interpreted vertically (as a social indicator), relating to different areal units, or horizontally (in which case a profile of an area is given). The cube can also be used to differentiate provision relating to different social or ethnic groups. The 'fact' X_{ij} can be one of many variables, relating to housing, education, social services provision, transport facilities, household type or recreation, although frequently such data are of restricted availability, and only census data are common.

sentation of the variations between places of particular 'goods' and 'bads':

> any square or cell in this matrix represents a *fact* in numerical form. For example, the cell marked X_{ij} represents how the people in a particular group (or perhaps the average of all people) in place or area j rate on criterion i; it could be the level of health in the Midlands region, or *per capita* income in the London Borough of Tower Hamlets. Any column in the matrix represents an *indicator* of how each area performs on the criterion in question. Any row shows a *profile* of the performance of the area in question on the various criteria of social well-being. A row back through the matrix, in the "who" dimension, would show the distribution of some attribute such as health or income among sub-divisions of the population in a specific area. (Smith, 1977b, p. 18)

In order that different distributions may be examined, Smith indicates how various patterns of expenditure and provision may be compared; his example compares education and armaments. Figure 2.3 illustrates a comparison of expenditure on two rather different types of local authority activity: road maintenance and primary schooling; the data are taken from 1978 sources, and relate to all non-metropolitan counties in England. The graph is an attempt to analyse the overall welfare achieved by mixing various types of expenditure; in this case, the returns are considered in terms of the variation in road accidents that follows from expenditure upon highways, and the numbers of primary places that can be provided within the local education authority. On the bottom left of the figure are represented the returns that could be gained following an expenditure of £20 per 1000 people; an authority can reduce accidents to 50 per kilometre of road (vertical axis), or can provide primary places for 47 children (horizontal axis). Naturally, any particular authority can spend its revenue according to its own priorities, or more strictly, the priorities of its voters and rate-payers. These are displayed in the form of an indifference curve, which represents a distribution that is satisfactory to a majority; typically, these are curvilinear, in so far as there will be some minimum values of any service required, and expenditure will never be zero. In this hypothetical case the line displaying the possibilities intersects the community's choice at a point that favours highway expenditure rather than school provision.

In reality, the community indifference curves are nearly vertical, indicating very little willingness to limit educational spending, and the line representing actual average expenditure bisects the indifference curve low on the highway spending axis (vertical), but high on the primary school axis (horizontal). Using this type of display,

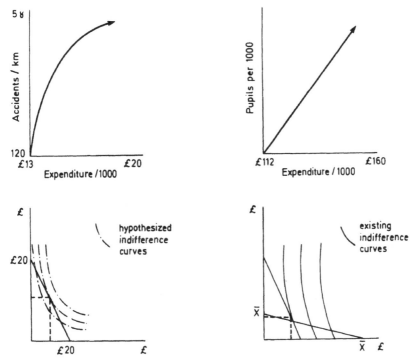

Figure 2.3   Variations in welfare with expenditure

The diagram is an adaption of Smith's discussion (1977a, p.47) and shows how a local authority may allocate funds to either road maintenance or education; in each case direct benefits result from higher expenditure: in the first instance from fewer accidents (top left), in the second from greater numbers of children in education (top right). Given a need to spend on both and with, say, only £20 per 1000 people available, it is likely that a local authority will spend in order to meet the community view, which is shown (bottom left) in the form of indifference curves. Analysis using English County data for 1977/8 suggests that the indifference curves are almost vertical, and that the mean level of expenditure ($\bar{x}$) is highly biased towards educational spending.

we are consequently able both to examine the impact of spending upon well-being (in this instance provision seems to be positively related to outcomes), and the priorities of the community; in this case, education is apparently more important than keeping death off the roads.

The mechanisms of deprivation

Whilst Smith's emphasis upon the welfare approach provides a means of measuring the outputs of particular mixes of provision, it leaves us to identify the actual processes that bring disadvantage about. We need to ask, for example, whether distributional problems are chance effects, or have more deliberate origins. We need to know whether resource provision is restricted to certain areas, or certain groups, and whether there exist problems of accessibility or allocation. A means of differentiating these issues is summarized in Table 2.2.

Table 2.2 A typology of deprivation, with examples

| | *Deprivation* | |
	Location-specific	*Class-specific*
In situ	National, regional health-service provision	Housing legislation
Spatial interaction	Local health-service provision	Local educational provision

Four separate ideas are isolated here. The most important is that deprivation may be restricted to particular groups (which implies that some aspatial process may be at work), or to particular areas, which may be a less malevolent issue. A subsidiary question is the notion of deprivation *in situ*, in the manner outlined by Peet, contrasted with deprivation that springs from mobility problems. Each of these cases will be examined in turn.

Figure 2.4 presents three contrasting maps of England, showing health provision and related indicators, measured at regional health authority scale. Figure 2.4a shows the distribution of capital expenditure on hospitals in 1971–2, which varies from £4259 per 1000 population down to £2556 (Noyce *et al.*, 1974). Their research has indicated that this variation is difficult to explain in rational terms (i.e. in relation to medical need) and that it seems to be statistically related to the distribution of high social class:

there are no regions of above-average spending which are not also high socio-economic status regions. Indeed, if one knew no other facts, it would be possible to explain two-thirds of the variation in community-health expenditure [and hospital expenditure – AK] by a knowledge of what proportion of the population in each region were managers, employers or professional workers. (1974, p. 556)

This variation is illustrated in Figure 2.4b, which shows the distribution of élite social groups.

The third map is included to suggest that resource variations may have impacts upon well-being, i.e. that deprivation is not a trivial problem. Clearly, it is impossible to prove a causal connection between hospital construction and mortality. It is, however, interesting that even when allowances are made for age and sex variations (the map shows 'the number of deaths in a region expressed as a percentage of the number that would have been expected if the age–sex specific death rates for England and Wales as a whole had been applied to the age–sex composition of the population in the region' (Office of Health Economics, 1977, p. 27), the most favoured regions in terms of provision have the lowest death rates.[3] The most neutral comment that could be made is that 'evidently as late as 1971–2 no effort was achieving success in directing new capital to deprived regions' (Noyce *et al.*, 1974).

HOSPITAL CAPITAL EXPENDITURE (1971 - 2)   SOCIAL CLASS DISTRIBUTION (1971)   MORTALITY RATIOS (STANDARDIZED) (1972)

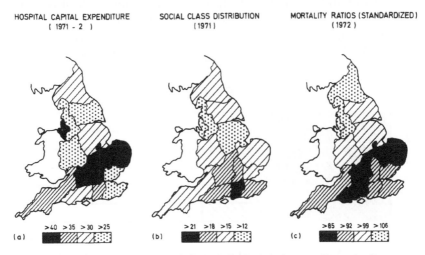

Figure 2.4   The distribution of (a) hospital capital expenditure in the regional health authorities 1971–2; (b) percentage of the population in socio-economic groups 1, 2, 3, 4 and 13: employers and managers; and (c) the standardized mortality ratios

The example chosen above is one in which all those in a particular region gain, or lose, with respect to medical provision; the fact that there exist differential proportions of high- and low-income groups within different regions does not alter this. Nor does a spatial interactional element arise; if there is any effect upon the health of the region's population, it is as a result of the overall level of expenditure, not the local distribution of hospitals. Indeed, mobility and movement need only concern us in relation to patients crossing into adjacent regions for the treatment they require (Buxton and Klein, 1975).

Of course, there are numerous facilities to which the consumer cannot expect to be transported (as is normal with many hospital users), and to which (s)he must make his or her own way. If we continue the medical theme, we can find a good deal of evidence that shows the extent to which certain locations are deprived in terms of basic provision, with the result that long journeys are required by the patient, who frequently earns this label. This is shown most starkly in rural areas where declining populations and settlement policies have brought about a concentration of resources at a small number of key points. Such strategies aim for economies of scale, whilst passing the costs of movement to the consumer. Even without the concomitant decline in public transport, this would represent location-specific deprivation, in that there would exist time and cost penalties upon certain locations. As it is, some villages are now in a very poor position *vis-à-vis* basic medical services such as dentists, who do not normally undertake house-calls. As Figure 2.5 shows, despite the possibility of organizing

Dentist

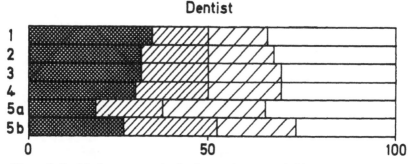

Figure 2.5   Various strategies for improving accessibility to dental services in East Anglia (from Moseley, 1979)

Moseley's study is an investigation into the problems of accessibility and mobility in rural areas. He uses six strategies of organization for bus and rail links in Norfolk, and notes the effects on – in this example – users of dental services. The bar graph shows the different proportions of the local population which are able to utilize dentists using different transport modes:

white:	access only possible using car
broad diagonal:	access by car and public transport
narrow diagonal:	access by public transport or on foot
stipple:	access impossible

Figure 2.6   Access to dental services in Newcastle upon Tyne, 1976

The maps on the right show the relative ease of obtaining dental treatment within the city; the hours worked by practitioners have been scaled by distance from school sites to surgeries. Figure 2.6(a) shows the scores for schoolchildren attending clinics run by the School Dental Service; Figure 2.6(b) shows the results of a potential model, based on the possibilities of schoolchildren using any dentist within the city's NHS system. In each case, the access scores are expressed as a percentage of the highest possible scores.

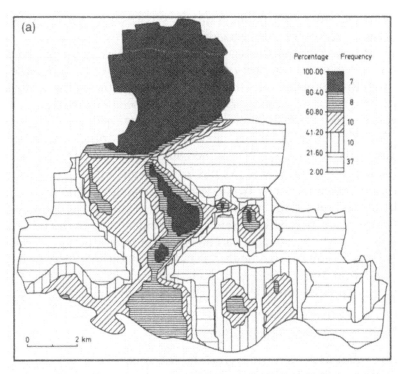

(a)

Percentage	Frequency
100·00	7
80·40	8
60·80	10
41·20	10
21·60	37
2·00	

0   2 km

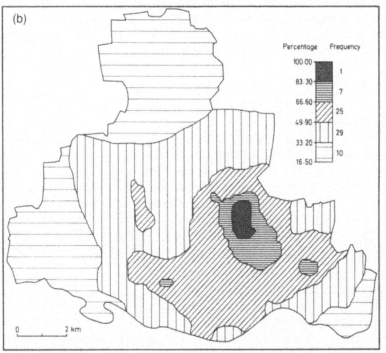

(b)

Percentage	Frequency
100·00	1
83·30	7
66·60	25
49·90	29
33·20	10
16·50	

0   2 km

various strategies for public transport services, the overall level of effective access for this particular part of East Anglia could not be raised above about 45 per cent of the population; some villages have in essence to rely on personal transport, and for households for whom this is unavailable, the time constraints of the public network (or its non-existence) suggests that about one quarter of the population cannot normally visit a dentist (Moseley, 1979).

Moseley rightly terms accessibility 'the rural challenge'. However, it should not be assumed that it is only the extreme cases that are important in terms of this form of deprivation. To continue this particular example, dental facilities are poorly distributed even throughout urban areas, due to the freedom enjoyed by dentists to locate where they wish. Figure 2.6 presents a map of the relative ease with which a dentist can be visited in Newcastle-upon-Tyne, and it is important to note that a study of schoolchildren in the city found that those areas with poor levels of access had also the poorest rates of dental hygiene (Bradley, Kirby and Taylor, 1978).

Examples of class-specific deprivation

The examples noted above are in line with recent debates concerning the malfunctioning of the spatial economy. Some of the problems faced both by particular regions, and by parts of cities (such as the inner areas) can be linked up with some aspect of resource deprivation: a poor infrastructure, or outmoded public (and private) facilities. In line with such thinking, aid can be concentrated into the particular areas via either regional policy or the various measures designed to aid the inner areas, such as the Inner City Partnerships.

Such thinking can frequently assume that spatial deprivation is just that, i.e. spatial in origin. There is a good deal of evidence, however, particularly at the smaller spatial scales, that such deprivation in terms of resources may be associated with particular social groups, who also happen to possess some spatial existence. The first example will serve to illustrate this.

The 1969 Housing Act introduced the General Improvement Area (GIA), usually a small group of streets with an essentially stable population that could benefit from the provision of home-improvement grants. Although the GIA is a local-authority administered activity, it is funded by the state; a local authority must consequently achieve the approval of the Department of the Environment before declaring such an area. Given the finite nature of financial resources, it is natural that some spatial search procedure is undertaken in order to find an area in which the intentions of 'employing limited resources in such a way as to bring about the maximum amount of improvement with the resources available' can be achieved. Clearly some neighbourhoods are too wealthy, or too young, to require housing aid; these excluded, we might, however, expect that the most needy districts would automatically receive consideration, on the basis of housing criteria alone.

In his study of the implementation of GIAs in Huddersfield, Duncan noted that such a rational approach was not in evidence (1974). As Figure 2.7 indicates, the selection process was a complicated affair, that in fact involved a series of considerations. Initially, unsuitable neighbourhoods were withdrawn, due to their satisfactory housing conditions, their relative youth or because they were not residential. This left a residue of areas in need of improvement, although from these, neighbourhoods where slum clearance rather than improvement was likely, were also removed.

In terms of the present discussion, the interesting decisions were made at the next step, wherein 36 potential areas were reduced to 8 possible sites. This was achieved by excluding areas with high proportions of Asian families and/or elderly households. Due to the vagaries of the housing market, such groups are clearly recognizable in space, and can be readily removed from the process of consideration on the grounds that they will be unlikely to take up the same level of improvement grants as other, more 'stable' areas.

In this instance, therefore, there is clear evidence of resource deprivation in some areas; indeed, the selective nature of the financial assistance in question makes this inevitable from the outset. However, we might reasonably expect that a GIA will be declared on the grounds of housing need, rather than the race or age of the houses' occupiers. The Huddersfield example implies that in this type of case, certain groups, because of their spatial existence, can be easily discriminated against, and that they may suffer deprivation in consequence. When we come in turn to link this finding back to our remarks about attempts to overcome deprivation, it becomes clear that spatial problems may not be simply that: i.e. accidental administrative malfunctions manifest in space. Instead, they may represent long traditions of deliberate decision-making directed against certain groups. This chapter is not the place to examine the various approaches to an understanding of those who make allocative decisions, but there exist several studies that suggest that local authorities as a whole, and the individual professionals that implement particular policies, are both liable to

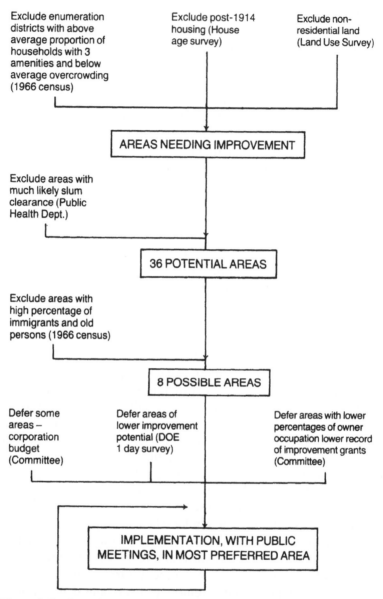

Exclude enumeration districts with above average proportion of households with 3 amenities and below average overcrowding (1966 census)

Exclude post-1914 housing (House age survey)

Exclude non-residential land (Land Use Survey)

AREAS NEEDING IMPROVEMENT

Exclude areas with much likely slum clearance (Public Health Dept.)

36 POTENTIAL AREAS

Exclude areas with high percentage of immigrants and old persons (1966 census)

8 POSSIBLE AREAS

Defer some areas – corporation budget (Committee)

Defer areas of lower improvement potential (DOE 1 day survey)

Defer areas with lower percentages of owner occupation lower record of improvement grants (Committee)

IMPLEMENTATION, WITH PUBLIC MEETINGS, IN MOST PREFERRED AREA

Figure 2.7   Steps in the decision-making process in the choice of a GIA site in Huddersfield (from Duncan, 1974)

The diagram illustrates the way in which various possible sites in Huddersfield were considered – and rejected – as being suitable for the declaration of a General Improvement Area.

subordinate the identification and overcoming of needs to other considerations (see, for example, Kirby, 1979c).

This suggestion may be underlined by examples of class-specific deprivation that involve movement. This said, it is very difficult to disentangle discrimination against particular classes or groups from policies that have implications for areas which happen to be co-incident with a particular group. It can easily be shown for example that in inner urban areas in the United States, blacks have to travel further in order to reach a physician than do whites (Shannon and Dever, 1974). This does not reflect, however, a spatial discri-mination against black ghettoes in the location of hospitals: indeed, there are far more hospitals in the core areas than in the suburbs. The former are thus locationally convenient, but financially inac-cessible to the majority of blacks, and consequently the long average journeys for care are accounted for by travel to what is (usually) a single *public* facility. In this particular case, therefore, the cause of the deprivation is rather more to do with structural inequalities than distributional ones, although the latter serve to add another layer of hardship to the other problems faced by urban blacks.

The types of examples that would unequivocally underline or undermine this theme are the issues of access that involve different groups being assigned to different public facility locations, or 'stations'. For example, do those who make use of Professional Employment Registers find that these stations are more accessible to them as a group, than are the more lowly Job Centres to those who have to make use of them? Similarly, we might ask whether there is any hard evidence that certain classes of offender are located in the country's peripheral prisons, whilst those who have committed types of 'middle-class crime' such as fraud, are located in open prisons near to centres of population: an academic question, except, of course, to those relatives incurring expense and hardship when making visits.

It is only in the field of education that these types of questions have been explicitly explored. Much recent research shows that many local education authorities use spatial districting methods to assign pupils from particular neighbourhoods to particular schools, with the result that the social mixes within different schools differ greatly. The extent to which this mixing is also related to the quality of educational provision is explored further in Chapter 3.

Summary

In this chapter the argument has concentrated upon the creation of deprivation as a phenomenon that may be very different from inequalities that develop in the marketplace. I have stressed that

Table 2.3 A comparison of the national classification (census data) with a classification of individual households (National Children's Bureau data)

Labels given to clusters	
Clusters unique to the spatial classification	*Clusters unique to the individual classifcation*
owner-occupier areas changing to furnished	small and single-parent families, owner-occupiers
very old owner-occupier suburbs	small and single-parent families, council
very new blue-collar owner-occupier estates	large families, overcrowded, owner-occupiers
mixed owner-occupier and unfurnished areas	overcrowded owner-occupiers, small number of rooms
mixed owner-occupier and unfurnished, ageing population	deprived large households in large council houses
mixed unfurnished and furnished, ageing population	deprived large households, owner-occupiers
unfurnished and skilled, ageing population	overcrowded, large households, owner-occupiers
council estates built for the old	
high-status furnished areas	
student areas	
reception areas (white)	
rural areas	
unfurnished, unskilled, no amenities	
walk to work, typical inner-city area	
black ghetto	
immigrants, unfurnished	
Types common to both classifications	
professional suburbs	blue-collar owner-occupier estates
non-manual suburbs	unfurnished artisan housing
executive housing estates	good council estates, recent
white-collar suburban estates	good council estates, established
	council estates with problems
	council estates with serious problems
	immigrant reception areas
	clearance areas

Source: after Cullingford and Openshaw (1979), Table 2.

types of deprivation develop as a result of distributional matters and that they are worthy of examination as independent – frequently spatial – issues. In the final analysis we must indicate the extent to which deprivation (as the term is used here) *is* a separate issue, and not merely a subordinate part of structural relations within society. However, it is very difficult to follow this line of thought through.

This is not to suggest that attempts have not been made to identify types of deprivation. Cullingford and Openshaw, for example, have examined two different data sets: one spatial (the Census of Population) and the other aspatial (data taken from the National Children's Bureau (Cullingford and Openshaw, 1979)). These they have submitted to cluster analysis, in order to test whether there exist unique geographical concentrations of deprivation, i.e. 'deprived places', in addition to 'deprived people'. They note distinct clusters: some are unique to the aspatial data, some unique to the geographical data, whilst some are common to both, as Table 2.3 outlines.

In the opinion of the researchers, all the unique clusters within the aspatial data are the result of aggregation effects, which swamp these mixes of variables when they are examined in a spatial context. Conversely, some of the unique clusters within the spatial data are seen to be the result of the census possessing a broader range of data than those of the National Children's Bureau.

Most interesting from our perspective is, however, the suggestion that there exist four types of 'spatial deprivation effects': these are italicized in Table 2.3. All four clusters are associated with inner urban areas, and could be taken as evidence that there are concentrations of deprivation arising from the juxtaposition of, for example, ethnic groups and poor housing. It must be stressed, however, that this kind of data analysis does *not* shed any light on our problem. In this instance, the concentration of ethnic groups in sections of the housing market (and thus in particular geographical locations) is a simple function of the 'normal' operation of the market; it is thus an economic, not a distributional question. As we have seen throughout this chapter, any empirical analysis of the latter issue must rest upon new sources of data, relating to education and health in particular.

Cullingford and Openshaw's research thus raises some interesting questions concerning the efficacy of spatially based policies, but cannot be taken as evidence either for or against distributional effects. For a fuller understanding of deprivation, we need to examine more specific issues, as the two following chapters indicate.

Notes

1 Peter Townsend's major study *Poverty in the United Kingdom* is the most recently published examination of deprivation; it is, however, not discussed here for two reasons. In the first instance, much of Townsend's discussion and analysis rests upon the notion of *relative deprivation*, with all the limitations that this immediately implies, and his attempts to quantify this measurement are not without critics (Piachaud, 1981, p. 420). More importantly, his study rests almost exclusively upon income, and thus on financial poverty; it does not, therefore, aid our investigation of deprivation as it is defined here.
2 This argument is of course reminiscent of that undertaken upon the value of studying managers of institutions like building societies and housing departments. For a résumé of the criticisms of this approach, see Kirby, 1979c.
3 Although capital expenditure has been used here, similar results emerge if hospital revenue and community-health expenditure are used. The arguments assume long-term stability of expenditure patterns.

References

Ambrose, P. (1977) 'Access and spatial inequality', in *Values, Relevance and Policy*, Milton Keynes, Open University.

Bassett, K. and Short, J. R. (1980) *Housing and Residential Structure*, London, Routledge & Kegan Paul.

Bradley, J. E., Kirby, A. M. and Taylor, P. J. (1978) 'Distance decay and dental decay', *Regional Studies*, 12 (5), 529–40.

Buxton, M. J. and Klein R. E. (1975) 'Distribution of hospital provision: policy themes and resources variations', *British Medical Journal*, 8, (February) 345–9.

Community Development Project (1975) 'Coventry and Hillfields: prosperity and the persistence of inequalities', *Final Report* (I), Coventry, CDP.

Cullingford, D. and Openshaw, S. (1979) 'Deprived places or deprived people?' *Discussion Paper 28*, Centre for Urban and Regional Development Studies, University of Newcastle-upon-Tyne.

Duncan, S. S. (1974) 'Cosmetic planning or social engineering?' *Area* 6 (4), 259–70.

Dunleavy, P. (1980) *Urban Political Analysis*, London, Macmillan.

Giddens, A. (1973) *The Class Structure of the Advanced Societies*. London, Hutchinson.

Hamnett, C. (1979) 'Area-based explanations: a critical appraisal', in Herbert, D. T. and Smith, D. M. (eds) *Social Problems and the City*, Oxford, Oxford University Press.

Hirschfield, A. (1978) 'Theoretical approaches to the study of urban deprivation'. *Working Paper 233*. Department of Geography, University of Leeds.

Holtermann, S. (1975) 'Census indicators of urban deprivation' *Working Note No. 6*, Census Research Unit, Department of the Environment, London.

Kirby, A. M. (1978) *The Inner City – Causes and Effects*. Corbridge, RPA Books.

Kirby, A. M. (1979a) 'Public resource allocation – spatial inputs and social outcomes', Ch. 16 in *Resources and Planning*, Oxford, Pergamon Press.

Kirby, A. M. (1979b) *Education, Health and Housing*, Farnborough, Saxon House.

Kirby, A. M. (1979c) 'Managerialism: a view of local authority housing', *Public Administration Bulletin*, 30, 47–60.

Miliband, R. (1973) *The State in Capitalist Society*, London, Quartet.

Moseley, M. (1979) *Accessibility – the Rural Challenge*, London, Methuen.

Noyce, J. Snaith, A. H. and Trickey, A. J. (1974) 'Regional variations in the allocation of financial resources to the community health services', *Lancet* 554–7.

Office of Health Economics (1977) 'The reorganised NHS', *Paper 58*, London, Office of Health Economics.

Pahl, R. (1978) 'Castells and collective consumption', *Sociology*, 12, 309–15.

Peet, R. (1975) 'Inequality and poverty: a marxist–geographic theory', Association of American Geographers, *Annals*, 65 (4), 564–71.

Piachaud, D. (1981) 'Peter Townsend and the Holy Grail', *New Society*, 57(982), 419–21.

Runciman, W. G. (1966) *Relative Deprivation and Social Justice*. London, Routledge & Kegan Paul.

Rutter, M. and Madge, N. (1976) *Cycles of Disadvantage*. London, Heinemann Educational Books.

Saunders, P. (1980) *Urban Politics*, Harmondsworth, Penguin.

Shannon, G. W. and Dever, G. E. A. (1974) *Health-care Delivery: Spatial Perspectives*, New York, McGraw-Hill.

Smith, D. M. (1977b) 'The welfare approach to human geography', in *Values Relevance and Policy*, Milton Keynes, Open University.

Thrift, N. (1977) 'An introduction to time geography', *CATMOG* 13, Norwich, Geo-Abstracts.

Townsend, P. (1979) *Poverty in the United Kingdom*, Harmondsworth, Penguin.

Weber, M. (1948) 'Class, status and party', in Gerth, H. H. and Mills, C. W. (eds) *From Max Weber: Essays in Sociology*, London, Routledge & Kegan Paul.

Part II

Inequalities

In Part I the emphasis was upon space, and the argument attempted to mount a justification for examining phenomena within a spatial perspective. Part II consists of two chapters which begin to examine the role of space in the operation of society in a more detailed manner.

Chapters 3 and 4 deal with three separate – but of course related – issues, namely education, health and the political system. I have considered them together because in each case, the activities (schooling, healing and voting) take place in a spatial domain. This is in itself unremarkable. More particularly, the argument emphasizes that the spatial organization of each activity produces basic inequalities between locations (in terms of educational provision, health care or enfranchisement), which may in turn constitute deprivation of the types outlined above.

One point is to be stressed at this juncture; I shall return to it again in Chapter 8, but will not reiterate it throughout the argument that follows in order that the point does not become strained. It is none the less of central importance. Simply, it is not to be assumed that spatial effects are to be treated as a 'fact' in themselves. To view things from a spatial perspective can be as meaningless as stating that the *Titanic* sank in the late afternoon, or that Hitler invaded Russia at five in the morning. In other words, spatial organization is only meaningful when related to the social processes that manipulate it: a constituency system is not the cause of malapportionment, but a means of describing its creation. To understand why gerrymandering, for example, occurs, we have to proceed to understand the deeper political processes at work within the state. Similarly, to focus upon school-districting procedures is only to examine the outcomes of various other relationships: between local education authorities and the state, between education officers and elected councillors, and between the local authority and the voters.

In short then, these chapters serve as a further argument for examining social issues on a spatial dimension; they are not to be taken as anything other than brief sketches of the issues, however.

3 Educational and medical provision

There are no such things as mistakes.

(Henry James: *The Europeans*)

This chapter deals with two themes that are superficially very different, namely health and education. They are of interest, and may realistically be considered together, however, because they each represent a social service that is designed (in the British case) to be universally available; in each case, however, spatial variations in provision occur, and it is the intention here to show that the scale of variation is such that educational attainment and personal health can be affected. We begin with education.

Education: a personal achievement?

Chapter 2 considered briefly the fact that material rewards are in part related to educational achievement, a phenomenon that is likely to become increasingly true as unemployment becomes concentrated in the unskilled and semi-skilled sectors of the workforce, due to accelerating industrial and technological change. This is in itself cause for concern, but more serious is the fact that poor educational attainment is traditionally inter-generational, as Table 3.1 suggests.[1]

There exist numerous explanations for this pattern of low achievement; King (1971) outlines four, namely the differential

Table 3.1 Variations in educational attainment

| Level of attainment | Father's occupation | | Ratio N-m/M |
	Non-manual	Manual	
	%	%	
O-level	79	45.5	1.74
A-level	32	8.9	3.70
Degree	12	1.5	8.00

Source: King, 1971.

ability; the differential provision; the differential access and the cultural discontinuity hypotheses. These represent in turn two basic tenets: that in essence attainment is either a function of personal factors (ability, or cultural attitudes), or public issues (the provision of, and ability to make use of, educational resources). Let us examine initially the personal factors.

A recent paper by a psychologist has explicitly examined what he terms 'the social ecology of intelligence', i.e. the extent to which the average intelligence quotient (IQ) exhibits spatial variations (Lynn, 1979). Drawing on four major studies, Lynn produces mean IQs for the standard regions of the British Isles, plus Eire; the results are displayed in Table 3.2.

Table 3.2 Regional variations in intelligence and aspects of attainment

Region	IQ	First-class honours degrees % of students
London-South East	102.1	5.32
Eastern	101.7	4.58
East and West Ridings	101.1	3.29
Southern	100.9	5.00
North Midland	100.8	3.55
North Western	100.3	3.83
Northern	99.7	2.73
South Western	99.6	4.39
Wales	98.4	3.34
Midlands	98.1	3.11
Scotland	97.3	3.92
Ulster	96.7	3.89
Eire	96.0	2.18

Source: extracted from Lynn (1979), Table 2.

The implications of this are clear. First, intelligence displays spatial variations, or, rather differently, that individuals from different regions are likely to perform differently when their IQs are measured. Using historical records dating back to 1751, Lynn argues that these variations have remained constant, and are the product of 'the transmission of phenotypic intelligence from one generation to the next', and migration between regions, which is 'selective for intelligence' (Lynn, 1979, p. 10); (cultural attitudes towards achievement are not considered). From this starting point the argument proceeds to suggest that this social ecology is manifest in different ways, such as personal achievement or unemployment; using principal components analysis, Lynn demonstrates that intelligence, the taking of first-class honours degrees, membership of the Royal Society, *per capita* income, unemployment and infant mortality rates are all statistically related (loading on the first, general component); (for a discussion of the statistical analysis in question, see Goddard and Kirby, 1976).

The problem with this type of spatial analysis is that it must remain speculative, due to the spatial scale at which the data have been analysed. A large literature exists to show that correlation coefficients are artificially inflated when data are aggregated; this is a function of the aggregation process, which averages out extreme observations and reduces the variance of both dependent and independent variables (see, for example, Taylor, 1977 for discussion on this point). This can be illustrated visually by Figure 3.1, which displays the distribution of first-class honours degrees at the

FIRST CLASS HONOURS   A-LEVELS   A-LEVELS (COUNTIES)

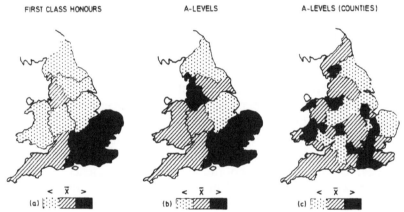

Figure 3.1 Variations in educational attainment, at different spatial scales: (a) (b) regions; (c) counties

regional scale, alongside additional measures of educational performance. If we consider in the first instance the relationship between intelligence at the aggregate scale and the attainment of first-class honours degrees, there is a statistical association, with a concentration of extremes in the east and west; this is reflected in a correlation of $r_s = 0.64$. Similar conclusions can be reached if A-level data are plotted; the figures are taken from Charlton *et al.* (1979). Once more, geographical concentrations are manifest, and a positive statistical relationship displayed ($r_s = 0.56$).

The dénouement comes when we consider the distribution of A-level candidates at the more detailed, county scale. From this, we can clearly see that the gross figures hide a great deal of local variation. Wales, for example, has overall an achievement rate that is slightly below the national average of about 18 per cent. However, local areas display figures that vary from 9.5 per cent through to 33.9 per cent; both examples are in fact the national extremes for England and Wales as a whole. This allows us to make two conclusions. The first is that the supposed link between intelligence and attainment may not be as close as Lynn's coarse study suggests; the second is that we require some explanation that accounts for large variations in attainment between neighbouring spatial units. It seems unlikely that IQ alone can account for this, because if we extrapolate the relationships found at the regional scale, we find that a county would have to possess a population with a mean IQ of about 112 in order to yield a one-third proportion of 18-year-olds sitting A-levels. This is, to say the least, improbable.

Attainment and expenditure

A recent study by sociologists Byrne, Williamson and Fletcher concludes that 'educational attainment is a public issue, not a private achievement, an artefact of the distribution of power in society rather than a distribution of intelligence' (1975, p. 46). Their research is noteworthy because it does not concentrate upon personal issues (differential ability or cultural discontinuity), but rather upon distributional questions; explicitly, the spatial distribution of financial expenditure upon, and the provision of, educational resources, measured in terms of the staff/pupil ratios, teacher salaries and *per capita* expenditure. They conclude that 'spatially defined variations in provision are strongly related to spatially defined variations in measures of socially significant educational attainment' (1975, p. 155).

In a major analysis of local education authorities (LEAs), they were able to use cluster analysis to show that specific patterns of expenditure and, consequently, performance exist. Six clusters were obtained, and these are sketched in Table 3.3. As the Table indicates, there appear to be clear distinctions between types of LEA on the bases of educational expenditure, the types of system operated (comprehensive or selective) and the outputs (expressed in terms of rates of staying-on after the school-leaving age and university entrance). Now clearly, the position being put forward here is a major simplification of the position taken by Byrne and his associates. They, for example, consider in detail various environmental factors, and the social-class composition of the different LEAs; using partial correlations, they suggest that these influences have a smaller impact on attainment than does provision. Nor has

Table 3.3 Local education authority clusters, 1970

Cluster profiles

Variables	1	2	3	4	5	6
Expenditure	Average	High	Average	Very high	Average	Very high
Attainment	Low	High	Average	High	Average	Very high
Politics	Labour	Conservative	Labour	Labour	Conservative	Labour
System	Selective	Selective	Comprehensive	Comprehensive	Selective	Mixed
Examples	Barnsley	Berkshire	Gateshead	Bristol and	Blackpool	Welsh
	South Shields	Solihull	Cardiff	Southampton	Leeds	counties
	Wigan	Cheshire	Doncaster	(only)	Most English	
				London	county	
				boroughs	boroughs	

Source: Byrne *et al*. (1975).

their work been without its critics, who are particularly worried about on the one hand the ecological level of analysis, and on the other the actual mechanisms by which inputs (cash) buy outputs (attainment); (Hutchinson, 1975; Pyle, 1976). The question of the detailed operation of the educational system *within* LEAs will be considered below; the present level of analysis can, however, be used to suggest an alternative to the social ecology of intelligence view of attainment. In other words, it seems reasonable to accept that the large variations that exist in terms of A-level entries in an area like Wales could be the product of the local attitude to educational expenditure, which as we have seen is particularly different from that found in other LEAs.

Provision within authorities

In a recent paper Williamson and Byrne have re-examined these issues in the urban setting. They observe that:

different groups of children, even when they are from similar socio-economic backgrounds, through living in different local authority areas or parts of towns, differ in their educational attainments . . . such a spatial structuring of educational inequalities, of which the phenomenon of inner-city deprivation is only the tip of the iceberg, is related to local education authority organisation, resource levels and politics, and is entrenched in the much larger system of local authority financing through the rate-support grant, which is financed by central government. (1979 p. 191)

These remarks introduce two issues: the mechanisms by which attainment is manipulated, and the nature of the system that allows, or perhaps encourages, these variations to exist. We may usefully begin with the role of the school itself (King, 1974).

In his examination of differential provision in schools, measured either in terms of equipment or teaching, King questions three separate assertions. The first is that working-class children receive poorer provision; the second is that attainment is correlated with 'material provision'; and the third is that 'the social class differences in educational attainment are (partly) explained by the differences in provision' (1974, p. 17). His study of the sixteen comprehensives within one LEA reveals some unexpected correlations, as Table 3.4 indicates. The table suggests that in line with other research, the proportions of children staying on after the minimum leaving age are inversely related to the proportions of children from manual backgrounds (social classes IV and V). Conversely, the levels of provision appear to be weighted in favour of schools with a high proportion of working-class children, in so far as pupil-teacher ratios are superior

Table 3.4 Attainment and provision in relation to high and low social class: comprehensive schools (correlations)

Social class	% 4th year staying on	% 4th year taking O-level	% Staff graduates	Pupil-teacher ratio
% I and II	0.24	0.24	0.16	0.37
% IV and V	−0.54	−0.28	−0.34	−0.69

Source: extracted from King (1974).

than for 'middle-class schools'. It is to be noted, however, that there is an imbalance in terms of teachers' skills, and that there is greater retentivity of staff in 'working-class schools': this may indicate poorer career prospects for poorer-qualified teachers, who consequently change their jobs less frequently.

Table 3.5 Attainment in relation to provision: comprehensive schools (correlations)

Attainment	Staff turnover	% Graduate staff	Pupil-teacher ratio	Material provision
% 4th year staying on	–0.43	0.44	0.75	–0.09
% 4th year with 5 or more O-level passes	−0.32	0.42	0.36	0.35

Source: extracted from King (1974).

When various aspects of provision are correlated with pupil attainment, King concludes that

the differences in the mean levels of pupil attainment in the schools are not in general associated with differences in their educational provision (such as language laboratories, gymnasia, libraries), but that low levels of attainment are associated with low proportions of full-time staff, high staff stability and low pupil-teacher ratios. (1974, p. 28)

From this analysis King goes on to suggest that low attainment amongst working-class children occurs despite superior provision, although once more he does not focus upon the relationship between the standard of teacher qualifications (numbers of graduate staff), the concentration of these teachers in the 'middle-class schools', and the higher attainment therein. Indeed, the use of partial correlation coefficients shows that the relationship between class and attainment falls away markedly if the proportion of graduate teachers is held constant.

These kinds of rather contradictory results have produced a general feeling amongst some educationalists that the individual school does not play an important role in the inter-generational process of attainment; expressed another way, many have argued that compensatory provision is misguided, in so far as 'education cannot compensate for society'.

An important new study shows that much of the research that has led to this conclusion has been inept, in the sense that it has concentrated upon non-scholastic attainment, and has ignored the variations in intake to schools (Rutter, Maughan, Mortimore and Ouston, 1979). Rutter *et al.* argue that children have differential abilities by the time of their arrival at school, and that these variations will remain; none the less, the school can and does affect the overall level of attainment; in other words, there will be a distribution of achievement, from good to bad, but a 'good school' will produce a higher minimum standard than will a 'poor school'. This can be illustrated by Figure 3.2, which assesses the impacts of different schools on their pupil intake. The graph incorporates the following information: first, the mean examination score (which is computed on the basis of one point for GCE passes, one half-point for good CSE passes) within a selection of twelve comprehensive schools, all located in the Inner London Education Authority; secondly, the intake within the school (assessed in terms of a standard Verbal Reasoning score, undertaken on entry into the school); thirdly, the distribution of examination scores across the different ability bands.

Figure 3.2 shows, therefore, two things. First, that examination success is related to initial abilities: in other words, the most able at the time of entry are also the most able at the time of leaving school. Secondly, and more importantly, examination success is determined by attendance at a particular school. Consider the distribution band 3; that is, the weakest group in terms of VR on intake. On average, these children achieve only one CSE pass. However, in the most successful school, Band 3 pupils achieve on average 1.25 passes – as good a rate of success as that managed by Band 1 children in the least successful school! In other words, a certain type of school can bring the weakest intake up to the standard achieved by a more advanced intake elsewhere. (It should perhaps also be stated that an emphasis upon formal qualifications here relates to the links that exist between educational attainment and employ-

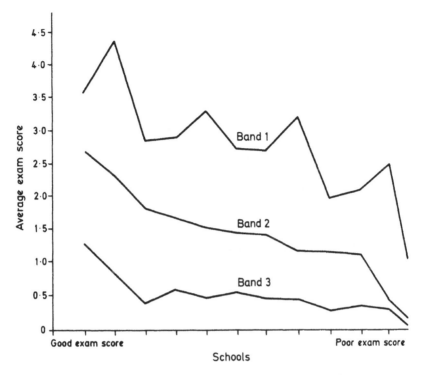

Figure 3.2   Attainment in 12 Inner London schools, by VRQ bands (source: Rutter *et al.*, 1979)

The diagram shows the mean examination score for each of the schools, disaggregated by performance bands: these represent the abilities of children on entry, and are measured using a verbal reasoning test. The schools have been arranged to show a consistent pattern, with 'high-attaining' schools on the left, and 'low-attaining' schools on the right.

ment; none the less, Rutter *et al.* also investigated aspects of pupil behaviour, and found these also to be a function of particular schools). In general terms, the school's impact appears to be a result, not of material provision, but rather of an overall attitude to academic success, discipline and achievement. Rutter *et al.* conclude their work by stating that 'schools can do much to foster good behaviour and attainments, and that even in a disadvantaged area, schools can be a force for the good' (1979, p. 205).

The spatial organization of education

The above analyses bring us up against a fundamental question. Clearly well-being, in the widest sense of the term, can be influenced by educational inputs; indeed, we can reasonably argue that social status itself is determined to some degree by the quality of schooling. What remains to be explained is how particular pupils are allocated to different schools – some good, some far from good.[2]

Secondary schools are organized to meet two constraints; on the one hand they must be relatively large in order that certain levels of provision (in science, or sport) may be maintained, whilst on the other hand they cannot be too large – one school per LEA would cause travel hardship for too many pupils. There is thus some spatial distribution of schools, coupled with some system of allocation. As far as the comprehensive sector is concerned (religious schools and fee-paying establishments draw on larger catchments), three systems co-exist nationally, as Dore and Flowerdew indicate.

Table 3.6 Types of spatial organization within the comprehensive sector in England and Wales, 1977

Allocation system	Proportion of LEAs
	%
Catchment area	51
(% involving some parental choice)	(19)
Feeder system	19
(% involving some parental choice)	(11)
Parental choice alone	27

Source: from Dore and Flowerdew, Table 1 (1978).

These figures are revealing, in that they suggest immediately that 57 per cent of schools are supplied with pupils as a result of parental choice. Because some parents are typically more mobile, knowledgeable and ambitious for their children, they will search out the schools with the greatest academic kudos, with the result that 'the schools with better reputations [are] effectively creaming many of their middle-class pupils from the remainder' (Dore and Flowerdew, 1978).

Although many LEAs permit individual choice within a particular framework, the majority of comprehensives do operate ostensibly on the basis of a formal allocation system. One in five uses a feeder system, an example of which is given in Kirby (1979b); in such cases, primary schools send on their pupils to one particular secondary school. The example quoted indicates that feeder systems are susceptible to manipulation, with some secondary schools drawing on concentrations of only one social group. One in two LEAs uses a catchment-area system, which may also produce imbalances within school populations. Dore and Flowerdew, for example, show how schools in Derby tend to draw only on local catchments, with the result that 'the boundaries tend to reinforce patterns of social segregation by allocating children to local comprehensives' (1978).

The existence of these types of system (and parental choice) means that access to the more successful secondary schools in an area may be constrained for particular households; given the social segregation that exists within cities, this pattern of constraint is likely to be cumulative, as Chapter 2 suggested, with class-specific deprivation emerging in consequence.

How should these imbalances be accounted for? Elsewhere, I have considered the different models that can be invoked to account for local-authority activities (Kirby, 1979d). In this instance we can immediately isolate factors such as party-political (ideological) attitudes to comprehensivization, the dictates of administrative efficiency, the needs of local capital (in terms of maintaining a diversity of education specializations and attainment), and the involvement of the state. In terms of education the latter has been particularly active in attempting to improve the standards of particular schools, via the inputs of positive discriminatory aid to inner-city areas. The existence of Educational Priority Areas explicitly recognizes that many inner-area primary schools (which, of course, draw on a local area, and do not employ feeder systems or permit wide degrees of parental choice) were decayed and poorly staffed (see, for example, Herbert, 1976). Criticism has been made of the EPA concept, suggesting that it represents a crude attempt to compensate for social tensions and inequalities, a criticism equally to be levied at other aspects of the inner-city programme, such as the Community Development Projects (Williamson and Byrne, 1979). Seen within the context of the present discussion, however, it simply represents an early acceptance of the fact that certain groups may be spatially constrained to utilize particular schools, and that the latter should not, as far as possible, be of a substandard nature.

Résumé

The present discussion is a very brief outline of the educational question, and has simplified several issues; in particular I have concentrated upon the provision of educational goods, and played down the role of the family (and the neighbourhood, a subsidiary spatial theme well discussed by Herbert, 1976). This is a distortion of the traditional wisdom, but it is clear that the sociologists of education, led by Byrne and Williamson, have rewritten the agendas of concern, and created enormous interest in the role of provision in the process of attainment. Their observations are worthy of quotation at length, in the sense that they reflect in their entirety the wider argument being presented in this work:

Pahl's exposition of the nature of the socio-spatial system is, like Rex's exposition of the nature of housing class, characteristically Weberian. Both systems refer to groups which exist in relation to structures of inequality. Both groups possess sufficiently similar characteristics, as a consequence of their structural relationships, to merit consideration as classes. The crucial point is that there is no necessary link to the economically derived structure of inequality, in the Marxist sense. Ray Pahl talks about classes defined by spatial constraints. John Rex talks about classes defined by administrative constraints. *Cohorts passing through schools in LEAs are defined by a combination of both* (my emphasis). (Byrne, Williamson and Fletcher, 1975, p. 154)

Spatial issues and medical concerns

It was suggested at the beginning of this chapter that education and medicine have much in common, in so far as provision is spatially organized, and may be inequitably distributed. This is true as far as it goes, but requires some elaboration. In particular, it is necessary to emphasize that both the *need* for medical care and the organization of provision have a spatial component. To a degree, this is true in the educational field, in that some spatial areas display concentrations of groups with particular educational needs: the children of one-parent families and ethnic minorities might be examples. In these cases, however, space is simply functioning as an accounting framework, and the concentration is a product of another factor, such as the housing market. As far as health is concerned, we have by way of contrast an example of an ecological process at work, i.e. local conditions may combine either to further the epidemiology of a particular disease, or to bring about some threat to personal health and safety.

The examples discussed in Chapter 5 are typical ones for showing the impact of externalities upon health. The long-term results of nuclear-energy leaks are to be imagined rather than demonstrated, but there have been numerous studies of the effects of conventional power stations and similar pollutants upon rates of respiratory disease (Wood and Lawrence, 1980). Traffic nodes are also dangerous, with carbon monoxide and lead poisoning being particularly serious for children living near to major roads. Airports are not noted for atmospheric pollution, but are responsible for high noise levels, which may bring about anxiety problems (see, for example, Giggs, 1979). Behind these peaks of pollution lie ambient levels which may in places constitute serious risks: concentrations of domestic smoke leading to winter smogs were, until recently, common features of urban areas, and studies such as those undertaken in London and Manchester indicate that there still exist relatively high concentrations of atmospheric pollution in some parts of some cities (Shepherd *et al.*, 1974).

This evidence should not, however, be taken to indicate that certain neighbourhoods are particularly dangerous, at least in the sense that this was the case in former times. Centuries ago urban areas as a whole were epicentres of disease (notably the plague); more recently large districts were susceptible to epidemics, such as that mapped by Dr John Snow in 1854. As this famous story indicates, ignorance of the mechanisms of disease allowed cholera (and other threats) to flourish in local water supplies, when simple remedial measures could have reduced the risks of outbreak (Hall, 1975, p. 27).

The fact that the risks faced in some urban neighbourhoods have markedly decreased does not mean that there is no longer any spatial variation in the ecology of disease. On the broad scale many epidemics display very similar patterns of diffusion: the spread of cholera in 1866 throughout the United States shows a good deal of similarity to the way in which influenza spread in 1918 (Pyle, 1979). Similarly, many unusual threats to health remain a minor problem simply because they remain endemic in a limited geographical area; Pyle, for example, examines Californian encephalitis and Rocky Mountain spotted fever, both of which possess names which allude to their concentrations in particular locations (Pyle, 1979, pp. 100–118).

Medical geography does not exist to identify peculiarities alone, however. A good deal of research is concentrated upon very common diseases which are not limited to particular areas, but which do vary markedly in intensity between localities; various types of cancer, for example, have been studied this way, in an attempt to isolate some independent variable that may account for the variations. As Pyle demonstrates in some detail, such spatial associations may be misleading. Variations in the incidence of measles outbreaks may be attributed to racial concentrations in American cities, or the related factor of the population density (1979, pp. 165–203). Producing the 'best' independent variable in such circumstances is not easy, and an incorrect interpretation may suggest incorrect policy outcomes. Giggs, for example, in his study of the spatial distribution of schizophrenia in Nottingham, noted a marked concentration of the mentally-ill in the inner areas of the city. This he accounted for by invoking an ecological explanation, i.e. the harsh environment of the inner city promoted anxiety and stress, an explanation that has also been put forward to account for the concentrations of suicides in similar locations (Giggs, 1973, 1979; Herbert, 1976b). This interpretation was rejected by Gudgin (1975), who emphasized in turn an intermediate variable of the sort already introduced above; he argued that schizophrenics were weak in both the labour and housing markets, and consequently tended to concentrate in cheap, inner-city housing. Yet a third consideration is introduced in turn by Dear, who emphasizes the relationship between the sick and the provision of mental-health care, which is frequently concentrated in inner areas because 'stable' suburban areas oppose the existence of such facilities: 'the ghettoization of patients . . . supports the general distance-decay hypothesis that patients in need of care will reside close to the source of care' (1977a, p. 592).

As far as this example is concerned, there is no definitive answer to the question as to which is the correct causal model. None the less, it usefully illustrates the problems of inference, and also reintroduces the role of provision in an understanding of health and well-being.

The provision of health care

In his review of health services, David Smith observes that 'each nation, region and community will have its own pattern of need or demand, the full satisfaction of which may require a unique spatial reponse in terms of the disposition of both fixed and spatial resources' (1979, p. 247). Clearly, it is very difficult to judge what an area's particular needs are, and to then estimate whether provision is satisfactory. It is, however, relatively easy to show that provision does vary markedly between spatial units, and that this variation is not in accord with an equitable distribution of resources (however that may be defined).

In the context of location-specific deprivation, the term was illustrated (in Chapter 2) with the example of expenditure upon hospitals and hospital equipment in the different regional health authorities (a situation that resulted in the Resource Allocation Working Party of the DHSS pointing to some authorities as 'over-financed'). This, however, is only the tip of an iceberg, as Julian Tudor Hart pointed out in his innovative analysis which coined the term 'the law of inverse care', and deals with the observed fact that 'from them that hath not, shall be taken away':

In areas with most sickness and death, general practitioners have more work, larger lists, less hospital support, and inherit more clinically ineffective traditions of consultation than in the healthiest areas; and hospital doctors shoulder heavier case-loads with less staff and equipment, more obsolete buildings, and suffer recurrent crises in the availability of beds and replacement staff. These trends can be summed up as the inverse care law: that the availability of good medical care tends to vary inversely with the need of the population served. (1971, p. 412)

Quite why this law works will be considered below; it should, however, be stressed that it can be applied at various spatial scales, and is certainly applicable at the international scale (Smith, 1979), as well as the national/regional example considered above. Variations at the urban scale are often more contradictory, due to the workings of central place principles: in other words, high-order functions such as hospitals tend to locate in inner areas, in order to serve centrally a large population. This often suggests an over-provision of resources in areas of deprivation, a denial of inverse care (see, once again, Smith, 1979, pp. 278–9). A similar example is provided in Figure 3.3, which outlines the provision of family-planning clinics in London, standardized against the numbers of women in each Borough 'at risk', i.e. of child-bearing age (Price and Cummings, 1977). The map shows clearly that there is good provision (in a comparative sense) in the Inner London boroughs; as the authors point out, however, much of this provision is of a centralized nature, serving specialized needs for larger populations, and the needs of women who work centrally during the day.

These kinds of considerations illustrate that principles of 'territorial justice' (provision to each area by need) are difficult to assess, let alone achieve (Pinch, 1980). The example also illustrates that provision to a particular area must be considered side by side with the issue of spatial accessibility; in this case relatively good provision for those in inner areas (whether designed for their benefit or not) is one side of the coin which can also mean long travel times and expense for those who live out in low-density suburban areas, and are not also working in the central area.

Increasingly, issues of spatial accessibility are coming to the attention of the medical geographical literature, and several examples of differential access are documented. Pioneer work was undertaken nearly two decades ago in Sweden, where spatial planning has to be considered in the context of a country with some very remote, sparsely settled regions. In this case hospital facilities

have been located in order that no individuals should be more than 8
hours from a particular facility (Godlund, 1961). Other countries
have adopted alternative strategies, using, for example, flying
doctors, although it is interesting to note that even attempts to bring
resources to the client cannot be totally successful, in so far as large
areas remain unsuitable for aircraft landings (see, for example,
McGlashan, 1972).

At a smaller spatial scale we might expect that accessibility issues
would recede in importance, but this is far from being the case.

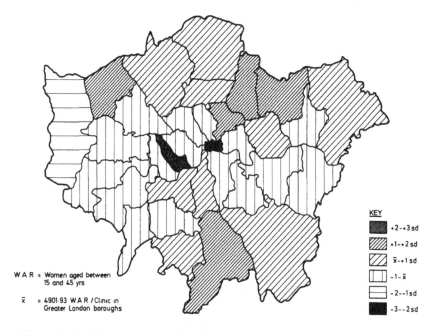

KEY

▨	+2-+3 sd
▥	+1-+2 sd
▧	x̄-+1 sd
☐	-1-x̄
▤	-2--1 sd
■	-3--2 sd

W A R = Women aged between
         15 and 45 yrs

x̄   = 4901·93 W A R / Clinic in
        Greater London boroughs

Figure 3.3   The provision of family-planning facilities in the London
boroughs (after Price and Cummings, 1977): data in standard deviations

Knox, for example, has examined the 'intraurban ecology of
primary medical care', concentrating upon Scottish cities (1978).
He notes a marked disparity in the distribution of community health
care: in Aberdeen, for example, 'over one-third of the surgeries and
nearly two-fifths of the doctors were located in the half-square mile
area . . . about half a mile to the west of the CBD. In contrast, most
of the suburbs are served badly by the location of GP's surgeries'
(1978, p. 421).

In common with many other cities, these suburbs contain large
local authority estates, many with relatively high concentrations of
low income. A similar picture is painted of Dundee (where two-

thirds of the ninety GPs are centrally located), although there is rather greater dispersal in Glasgow and Edinburgh.

Figure 3.4 illustrates the accessibility afforded to the different areas within Aberdeen *vis-à-vis* primary health care. It has been derived by Knox, using data on 'the actual fall-off in the registration of patients with distance from surgeries'; and by taking into account the distribution of public transport and private vehicles, he suggests that it can be interpreted as a broad measure of medical care.

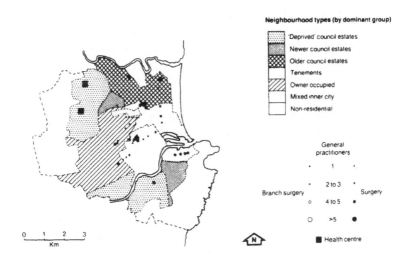

Figure 3.4   The distribution of general practitioners' surgeries in Aberdeen in 1973 (source: Knox, 1979)

Knox outlines the social ecology of medical provision, i.e. the distribution of GP's surgeries is superimposed upon a simple profile of residential areas. The concentration of surgeries in the inner urban area is quite marked, as are the virtual absence of GPs on some of the outlying peripheral estates. Knox's research into other Scottish cities, and investigations in other urban areas, such as Reading, reveal very similar patterns (Kirby and Scott-Samuel, 1981)

Drawing on additional maps of the kind displayed here, Knox picks out the high provision in the inner-city areas, but adds that 'beyond these core areas, however, accessibility appears to be correlated inversely with socio-economic status' (1978, p. 426). Discussing his analysis of all four cities, Knox concludes that

the examples of surgery location examined here have revealed disparities in medical care which are disturbing, not only because of the consequent inequalities in a supposedly egalitarian health service, but also because these disparities compound many of the existing spatial patterns of socio-economic disadvantage. (1978, p. 430)

Interestingly, very similar conclusions are reached in a study of primary care in New Zealand, where the authors noted that 'certain communities appear to be under-served by general practitioners . . . this situation is particularly critical in the newer, more youthful and lower-status suburban areas' (Barnett and Newton, 1977, p. 66).

Quite why these variations come about is difficult to determine, and several alternative explanations exist. First, it is perhaps worth pointing out that to the administrator, the patterns of distribution may appear 'rational'; on the one hand, inaccessibility deters utilization, as we shall see below, whilst in addition different social groups also make differential use of facilities. Consequently, low-status peripheral estates may actually have *hidden* needs that are not expressed in effective demand for services, and there may exist, therefore, little official awareness of the necessity for additional provision.

Other explanations concentrate upon the locational decisions taken by practitioners. Barnett and Newton isolate three goals possessed by individual physicians, namely a wish to maximize income, to increase social prestige, and to achieve high professional interaction. In such a context a doctor will attempt to locate where the neighbourhood is pleasant, where there is a large number of patients, and where colleagues are concentrated in back-up services (such as hospitals). The last two motivations account for concentrations in central districts of cities, whilst the wish to maximize social prestige may express itself in a suburban location (or a more bouyant region altogether, for that matter). Indeed, Knox identifies home-based surgeries in high-status suburbs as a fairly important determinant of inequitable distributions (Barnett and Newton, 1977; Knox, 1978).

Of particular interest here is the communality displayed by practitioners in both public and private health systems; Busch and Dale for example observe in relation to American doctors that 'on the whole, physicians tend to be located in more prosperous areas where there are supportive medical facilities' (1978, p. 174). Diesker and Chappel noted that practitioners were interested in hospital standards, acceptability of the location to their wives, and the 'openness' of the local medical community (1976).

When we consider the location of other medical services, different locational principles assert themselves. Dear has written extensively on the problems facing the mentally-ill, one of which is the location of the supportive services. Frequently, economies of scale dictate a concentration of health-care facilities, although such agglomerations then tend to attract residential opposition, as they

Table 3.7 Anticipated effects of a mental-health facility, and likely responses, by proximity: Toronto

(a) Likely impact	Expected direction of change: % of respondents		
	Minus	No change	Plus
Property values	46 (42)	45 (54)	8 (4)
Traffic flow	43 (19)	52 (76)	4 (5)
Resident satisfaction	39 (35)	39 (49)	22 (16)
Residential propensity to move	37 (28)	46 (54)	18 (18)
Neighbourhood image	32 (32)	48 (45)	20 (23)
Noise levels	31 (22)	58 (65)	12 (13)
Neighbourhood quality	28 (25)	52 (52)	20 (22)
Residential character	29 (23)	48 (53)	24 (25)
Attraction of undesirable residents	27 (23)	61 (69)	12 (9)
Personal safety	25 (21)	60 (65)	15 (15)
Property taxes	23 (13)	64 (72)	13 (15)
Visual appearance	20 (19)	63 (66)	17 (15)

Sample = 1090; figures in brackets relate to respondents already aware of a local facility (<131 respondents)

(b) Likely response	Proximity of facility		
	7–12 blocks	2–6 blocks	<1 block
Do nothing	39 (27)	31 (25)	30 (28)
Write to newspaper	1 (0)	2 (0)	2 (0)
Contact local politician	8 (13)	8 (5)	6 (3)
Contact local official	4 (0)	6 (5)	4 (6)
Sign petition	23 (47)	20 (25)	12 (17)
Attend meeting	10 (7)	16 (20)	14 (17)
Join protest group	5 (0)	6 (10)	6 (6)
Form protest group	2 (0)	2 (0)	4 (6)
Consider moving	8 (7)	(10)	22 (17)
Sample*	128 (15)	225 (20)	404 (35)

* Data in brackets: already aware of local facility

Source: from Tables 2 and 6, Dear *et al.* (1980).

constitute an unacceptable externality: 'in Philadelphia, for example, the proliferation of outreach facilities, dormitories, parking lots and street signs had caused one area to take on the appearance of an "outpatient ward" for the whole city' (Dear, 1978, p. 104). The net result of these two tendencies is that facilities, in this field at least, are not only urban (as opposed to rural or suburban) but centralized: 'such locations tend to proliferate in transient, rental accommodation areas of inner cities'. As already noted, this reflects the political abilities of suburban residents. Dear, Taylor and Hall, for example, noted in a city-wide study of Toronto residents that quite large proportions were relatively hostile to community-based facilities (Dear *et al.*, 1980). Moreover, the greater the hypothetical proximity to the respondent's home, the greater became the likelihood that hostility might evolve into political protest of the sort discussed in Chapter 1; as they also note, however, the figures relate to potential, rather than actual political involvement concerning this kind of issue (Table 3.7).

Mobility and morbidity

So far we have considered the spatial variations that exist in disease rates, and the spatial variations that occur with respect to medical provision. It now remains to consider whether these two issues are in any way related.

At the extremes there is no question that very small variations in provision can bring about major impacts: 'in poor nations, and certain poor parts of rich nations, human life chances are severely curtailed by the limited availability of basic health care' (Smith, 1979, p. 246).

At smaller spatial scales there is clearly a greater degree of uniformity in terms of health and ill-health, although rates of, for example, infant mortality do vary significantly within single cities. Indeed, Table 3.8 shows that this health indicator varies far more widely even than unemployment in the case of Liverpool; (the city's neighbourhoods have been collapsed into seven homogenous social

Table 3.8 Infant mortality and other indicators, Liverpool Area Health Authority social areas. Ratio data: city average = 1.0

Indicator	Social area						
	1	2	3	4	5	6	7
Infant mortality	1.51	0.28	0.81	0.33	0.37	4.00	1.01
Unemployment	2.29	0.15	0.30	1.27	0.70	0.88	1.68
Illegitimacy	2.04	0.13	0.23	1.43	0.66	1.01	1.02
Higher-education grants awarded	0.37	2.70	2.65	0.65	0.81	0.72	0.94

Source: Kirby and Scott-Samuel (1981).

areas). This is a function to some degree of localized stress (such as housing conditions), but also a function of the use made of health services. The explanation for this pattern of use is, however, a complicated one.

Initially, we can trace a good deal of evidence that different social classes utilize even free health services to different degrees. In the preventive sector, for example, cervical cancer rates are nearly six times higher for social class 5 than for class 1, reflecting the variable rates of screening:

Table 3.9 Class and the use of health care

Variables	Social class 1	Social class 5
% of women in the North West	2.6	8.3
% of women being screened, 1968	7.7	4.1
% of Salisbury, dental visit in past 6 months	35.0	9.0
% using radiography clinics	29.4	24.3

Source: compiled from Waddington (1977).

As Waddington observes, 'it is of course true that one cannot compel people to use health services', although as he continues to demonstrate, plenty can be done to remove taboos and prejudices from particular groups and to improve communications (perhaps by simply making, for example, smear tests administratively easier to obtain). There is, however, the spatial question: a *mobile* health clinic based in Southwark reveals a pattern of use that is basically in line with the class distribution in the area:

Table 3.10 Use of mobile clinic by social class, compared with the underlying social mix

Social class	Clinic attendance %	Southwark population %
1	1.2	1.6
2	17.2	7.7
3	42.4	52.6
4	18.5	21.9
5	16.3	16.2

Source: Table 4, Waddington (1977).

Is this revealing? The literature on the spatial factor in resource use is thin, but a case seems to be made that where resource provision is uneven, less mobile groups are deterred from travelling to the nearest location, even though they might utilize the facility if it were available, as with the peripatetic clinic above.[3]

Once again we may begin with the extreme cases: in developing countries, for example, a good deal of research suggests that potential users 'have to go to extraordinary lengths – travelling long distances to facilities' (Bicknell and Walsh, 1976). Once more, a solution to the under-use of resources such as family-planning clinics may lie in the improvement of spatial access. The distance factor has been seen as an important obstacle to the speedy diffusion of birth control in Latin America; Fuller notes, for example, that 'the distance variable emerges as the single most powerful discriminator between users and non-users of contraceptive techniques' (1974, p. 331). The extent of this distance-decay effect is emphasized by a wider study from Africa (Jolly and King, 1966; Smith, 1977a). From this we can conclude that attendance falls in frequency as accessibility declines, although out-patients are prepared to travel longer distances for hospital treatment than for first aid. This may in itself be revealing, and could indicate that the frictions of distance cause prospective patients to avoid preventive medicine, and to use facilities only when serious illness necessitates remedial (hospital) care.

In some societies distance and travelling costs are merely one part of the overall cost of using medical care. The problems facing the American poor, particularly urban blacks, have been extensively examined, and it has been noted that blacks, who live in general in proximity to some of the world's finest clinics and hospitals, have extremely poor access to health care on cost grounds: 'to the poor inhabitant of the inner city . . . the nearby hospital might just as well be in Uganda in so far as real accessibility is concerned' (Smith, 1977a, p. 313). As a result blacks are likely to make less use of general practitioners and more use of remedial care when illness necessitates (Earickson, 1970). As the final irony, blacks may also have to travel long distances to make use of the isolated hospitals that offer free treatment.

In Britain the financial imperative is absent, and consequently it is meaningful to ask patients about their preferences for particular practitioners, and their ability and willingness to travel to different surgeries and clinics. In a study in West Glamorgan, for example, Phillips questioned respondents on three sets of issues: their mobility, their attitudes to the administrative constraints exercised by different doctors, and the performance of the latter (1979). In general, respondents from different areas and different social classes showed a good deal of satisfaction with their GPs, agreeing, for example, that their doctor was approachable, although they also tended to agree with suggestions that health centres should be scrapped. When presented with the statement 'it is a difficult journey to get to the doctor's surgery'; 'if I thought I could get on the list of a more conveniently located practice, I would change'; and 'if I really need to visit the surgery, the travelling does not discourage me from attending', respondents produced fairly consistent replies, although 'lower-status' individuals tended to emphasize these problems of access more than those in the more mobile, higher-income groups. Phillips notes that 'low-status repondents claimed a greater willingness to change their doctors for the sake of convenience'; he continues that they were also 'more likely to be discouraged from attending at surgery because of the travelling involved', and concludes that

such a finding does indeed reflect a social 'inverse-care' law whereby, although having access to the same broad range of GP services, low-status persons find themselves placed at a greater disadvantage than do high-status persons and hence less well able to use the services effectively. (Phillips, 1979, p. 821)

An overview of space

These studies suggest that the problems of accessibility may constitute a link between the poor distribution of resources in some areas, and the spatial variations that also exist in terms of morbidity and mortality. Curiously, no studies have followed up these outlines in order to examine the results, in terms of patients' well-being, of inequitable resource provision. The only analysis known to the author that attempts this link is an investigation of dental care provision and dental health in Newcastle upon Tyne (Bradley, Kirby and Taylor, 1978; Kirby, 1979b). As Ambrose observes, . 'the research suggests that poor dental health might be as much a question of accessibility as of attitude and that, if so, policies aimed at dispersing practices more evenly throughout the city might be more effective than policies aimed at changing people's attitudes' (1977, p. 107).

How then, can we summarize the spatial role in health care? Dear suggests five considerations, namely 'location as physical distance'; 'location as catchment'; 'location as social distance'; 'location as externality', and 'relative location' (1977a; 1978). Location as externality has already been examined, and relates to the placing of medical facilities away from 'sensitive' neighbourhoods; in other words, resources may not be located in relation to need, but rather where influential public opinion permits. Location as physical distance has also been considered, and although Dear notes little distance-decay effect in the field of mental-health provision in the USA, evidence from less mobile societies has been presented above; (see also 'relative location', below).

Dear allows us to consider three other spatial issues, the first of which is the catchment system. As with schools, catchments can distort accessibility patterns, as Taylor shows rather well in his analysis of provision in Anglesey (1977, p. 308). Secondly, he introduces the notion of social distance, which may be taken to mean several different things: it may refer to the process whereby previous users of a facility recommend it to a non-user, for example, in which case it becomes 'closer' in social space, if not in real terms. A less esoteric example is that of hospital discharges, where there is some evidence that patients living close to a hospital will be discharged more readily than those living further away from back-up care. The final example considers relative location, suggesting that in some instances, users may not opt for the nearest facility, but may consider the full spatial range of options. As far as mental-health provision is concerned, Dear notes that anonymity may encourage some patients to travel long distances to seek treatment; the same may well be true for clinics offering family-planning or venereal-disease consultations, for example.

Summary and conclusions

The aim of this chapter has been to underline the spatial issues that exist in the consumption of public goods like schools and clinics. To reiterate, it is not being suggested that an optimization of the spatial system of delivery would remove all the location-specific deprivation within education or health-care provision; there would still exist inequalities in terms of individual attitudes to education or long-standing fears of medical matters, not to mention waiting lists or inabilities to pay. None the less, it seems clear that individual schools are 'better' than others, and that the spatial organization controls access to these institutions. Equally, health-care facilities are unequally distributed, whilst some areas are, medically, more dangerous than others. The net result of these processes is that location may, again, determine social status, and in a very real sense, *where* you are dictates *what* you are.

Notes

1 More recent data continue to suggest major imbalances in, for example university entrance, which shows a manual/non-manual ratio of 21:79.
2 Although this section continues to concentrate upon the British case, the spatial organization of education has also had important implications in the American context. A long series of constitutional battles was fought over the rights of states to maintain racial concentrations within particular schools, in order to achieve racial desegregation within the education system as a whole. This is discussed by Taylor (1977), who also gives details of the way in which racial balances can be achieved between different schools, using linear programming methods.
3 It is perhaps indicative that in rural areas in Britain, patients living more than 1 mile from a dispensing chemist have traditionally been able to have their drugs supplied directly by the general practitioner; approximately 5% of prescriptions are dispensed this way.

References

Ambrose, P. (1977) 'Access and spatial inequality', in *Values, Relevance and Policy*, Milton Keynes, Open University.

Barnett, J. R. and Newton, P. (1977) 'Intra-urban disparities in the provision of primary health-care; an examination of three New Zealand urban areas', *Australian and New Zealand Journal of Sociology*, 13(1), 60–8.

Bicknell, F. O. and Walsh, D. C. (1976) 'Motivation and family planning: incentives and disincentives in the delivery system', *Social Science and Medicine*, 10 (11/12), 579–84.

Bradley, J. E; Kirby, A. M. and Taylor, P. J. (1978) 'Distance decay and dental decay', *Regional Studies*, 12(5), 529–40.

Busch, L. and Dale, C. (1978) 'The changing distribution of physicians' *Socio-economic Planning Sciences*, 12, 167–76.

Byrne, D., Williamson, B. and Fletcher, B. (1975) *The Poverty of Education*, Oxford, Robertson.

Charlton, W. A., Rawstron, E. M. and Rees, F. E. H. (1979) 'Regional disparities at A-level', *Geography*, 64, 26–36.

Dear, M. (1977a) 'Psychiatric patients and the inner city', Association of American Geographers, *Annals*, 67(4), 588–94.

Dear, M. (1978) 'Planning for mental health care: a reconsideration of public facility location theory', *International Regional Science Review*, 3(2), 92–111.

Dear, M., Taylor, S. M. and Hall, G. B. (1980) 'External effects of mental-health facilities', Association of American Geographers, *Annals*, 70(3), 342–52.

Diesker, R. A. and Chappel, J. A. (1976) 'Relative importance of variables in determination of practice location: a pilot study', *Social Science and Medicine*, 10(11/12), 599–64.

Dore, C. and Flowerdew, R. (1978) 'Allocation to comprehensive schools', mimeo, Department of Geography, University of Lancaster; published in *Manchester Geographer* (1981).

Earickson, R. (1970) 'The spatial behaviour of hospital patients', *Research Mimeograph 124*, Department of Geography, University of Chicago.

Fuller, G. (1974) 'On the spatial diffusion of fertility decline: the distance to clinic variable in a Chilean community', *Economic Geography*, 50(4), 324–34.

Giggs, J. (1973) 'The distribution of schizophrenics in Nottingham', Institute of British Geographers, *Transactions*, 59, 55–76.

Giggs, J. (1979) 'Human health problems in urban areas', in Herbert, D. T. and Smith, D. M. (eds) *Social Problems and the City*, Oxford, University Press.

Godlund, J. (1961) 'Population, regional hospitals, transport facilities and regions: planning the location of regional hospitals in Sweden', *Lund Studies in Geography*, B21.

Gudgin, G. (1975) 'The distribution of schizophrenics in Nottingham – a comment', Institute of British Geographers, *Transactions*, 64, 148–9.

Goddard, J. B. and Kirby, A. M. (1976) 'An introduction to factor analysis', *CATMOG*, 7, Norwich, Geo-Abstracts.

Hall, P. G. (1975) *Urban and Regional Planning*, Harmondsworth, Penguin.

Herbert, D. T. (1976a) 'Urban education: problems and policies', in Herbert D. T. and Johnston, R. J. (eds) *Social Areas in Cities*, Chichester, Wiley.

Herbert, D. T. (1976b) 'Social deviance in the city: a spatial perspective' in Herbert D. T. and Johnston, R. J. (eds) *Social Areas in Cities*, Chichester, Wiley.

Hutchinson, D. (1975) 'Areas of difference: a critique of the work of Byrne and Williamson on regional inequalities in educational attainment', *Quality and Quantity*, 9, 171–83.

Jolly, R. and King, M. (1966) 'The organisation of health services' in King, M. (ed.) *Medical Care in Developing Countries*, Oxford, University Press.

King, R. (1971) 'Unequal access in education – sex and social class' *Social and Economic Administration*, 5(3), 167–75.

King, R. (1974) 'Social class, educational attainment, and provision: an LEA study' *Policy and Politics*, 3(1), 17–35.

Kirby, A. M. (1979b) *Education, Health and Housing*, Farnborough, Saxon House.

Kirby, A. M. (1979d) 'Towards an understanding of the local state', *Geographical Paper 70*, Department of Geography, University of Reading.

Kirby, A. M. and Scott-Samuel, A. (1981) 'Health and health-care in the inner city' *Reading Geographer*, 8, pp. 31–42.

Knox, P. L. (1978) 'The intraurban ecology of primary medical care: patterns of accessibility and their policy implications', *Environment and Planning*, A 10, 415–34.

Knox, P. L. (1979) 'The accessibility of primary care to urban patients', *Journal of the Royal College of General Practitioners*, 29, 160–8.

Lynn, R. (1979) 'The social ecology of intelligence in the British Isles'. *British Journal of Social and Clinical Psychology*, 18, 1–12.

McGlashan, N. (1972) *Medical Geography*, London, Methuen.

Phillips, D. R. (1979) 'Public attitudes to general practitioners' services', *Environment and Planning*, A, 11, 815–24.

Pinch, S. (1980) 'Local authority provision for the elderly – an overview and case study of London' in Herbert, D. T. and Johnston R. J. (eds) *Geography and the Urban Environment*, 3, Chichester, Wiley.

Price, D. G. and Cummings, A. J. (1977) 'Family planning clinics in London', *Working Paper* 2, Department of Geography, Polytechnic of Central London.

Pyle, D. (1976) 'Aspects of resource allocation by local education authorities', *Social and Economic Administration*, 10 (2), 106–21.

Pyle, G. (1979) *Applied Medical Geography*, Washington, Winston.

Rutter, M., Maughan, B., Mortimore, P. and Ouston, J. (1979) *15000 Hours*, London, Open Books.

Shepherd, J., Lee, T. R. and Westaway, E. J. (1974) *A Social Atlas of London*, Oxford University Press.

Smith, D. M. (1977a) *Human Geography – a Welfare Approach*, London, Edward Arnold.

Smith, D. M. (1979) *Where the Grass is Greener*, Harmondsworth, Penguin.

Taylor, P. J. (1977) *Quantitative Methods in Geography*, Boston, Houghton-Mifflin.

Tudor Hart, J. (1971) 'The inverse care law', *Lancet*, 27 February, 405–12.

Waddington, I. (1977) 'The relationship between social class and the use of health services in Britain', *Journal of Advanced Nursing*, 2, 609–19.

Williamson, B. and Byrne, D. (1979) 'Educational disadvantage in an urban setting', in Herbert, D. T. and Smith, D. M. (eds) *Social Problems and the City*, Oxford, University Press.

Wood, C. and Lawrence, M. (1980) 'Air pollution and human health in Greater Manchester', *Environment and Planning*, A12, 1427–39.

4 Electoral organization

The British electorate was being fobbed off with what Plato called 'the noble lie'.

(Richard Crossman: *The Crossman Diaries*)

Elections in three dimensions

If we were to prepare a checklist of phenomena that are commonly thought to constitute a democratic political system, the attributes of, in Miliband's words, 'universal suffrage [and] free and regular elections' would undoubtedly emerge near the top of the list. Elections are the most visible and tangible aspects of the participative rights of the individual. Even in countries where citizens are denied fundamental freedoms, and where political parties (except for that which rules) are outlawed, elections take place regularly.

This chapter deals with an aspect of the political process that has frequently been underplayed, namely the fact that an election is not a single ideological struggle, but a simultaneous contest in which hundreds of candidates attempt to win the right to represent many small spatial areas. Of those nations thought of as 'democratic', only Israel does not display a constituency system, for there the phrase 'the national interest' takes on a new dimension. Elsewhere, local issues are seen as important and worthy of expression in local political contests.

In itself this is not remarkable: the relationship between a political representative and his or her constituents is an important check upon the former's behaviour, and a useful sounding-board for the latter's needs and demands. However, the spatial factor also has a malevolent aspect, namely that the existence of small areal units invariably leads to types of spatial bias; expressed simply, votes are worth more in some constituencies than in others. This may be a chance occurrence in some instances; in others, it is the result of deliberate engineering. Below, we examine how these types of bias may constitute location-specific and class-specific deprivation.

Spatial and electoral bias

It is election night, and the results have yet to come in. The pundits are re-examining the opinion polls, but are sceptical of correctly translating the likely distribution of votes into predictions of seats won and lost. This reticence is, however, a product of experience, and any commentator does well to remember the fortunes of the different parties in various British elections.

Table 4.1 Votes gained and seats won, British general elections

Date	Party	% Share of votes	% Share of seats
1906	Conservative	44	23.5
1910	Liberal	44	40.0
1964	Labour	44	50.2

As Table 4.1 indicates, the achievement of a certain level of popularity with the nation's voters does not guarantee a particular level of political power. Indeed, the Liberal landslide of 1906 saw the Conservatives still gaining 44 per cent of the national vote, whilst a similar level of support took Harold Wilson into office in 1964. Nor is this a British phenomenon:

Table 4.2 Minimum percentage of the population that could elect a majority in US state legislatures

	Senate		House of Representatives	
State	1962	1968	1962	1968
Nevada	8	50	35	48
California	11	49	45	49
Florida	12	51	12	50
Vermont	47	49	12	49

Source: extracted from Taylor and Johnston (1979).

Table 4.2 picks out the extreme cases, but none the less shows that until steps were taken to overhaul the system in the 1960s, tiny proportions of the vote could capture the state assemblies. Let us examine in turn how this may occur, beginning with the existence of different types of electoral bias.

A good deal of research effort has been expended by political scientists upon the correlation that exists between votes (V) and seats gained (S), and it has been noted that in many instances the relationship between the two can be expressed as follows: (A and B represent the two main parties fighting the election).

$$\left(\frac{VA}{VB}\right)^3 = \frac{SA}{SB}$$

Expressed verbally, this indicates that the ratio of votes cast for party A and party B will be equal to the cube root of the ratio of the number of seats that they actually win. Thus, if in terms of votes

A = 47%   B = 53% (ratio 1.12:1)

then we expect, in terms of seats

A = 41%   B = 59% (ratio 1.43:1)

This 'cube law' does not hold universally, in either time or space, but isolates an important feature of many electoral systems, namely the existence of a *'winner's bias'*. It comes about because the spread of support for parties tends to be normally distributed, i.e. most constituencies are fairly evenly balanced, and only a minority is heavily inclined to one party. This means that a very slight swing to one party or another captures an inordinate number of seats (a simplistic analogy would be to compare an election with trench warfare, in which an enormous number of troops are concentrated in a very small area: this means that every time there is a slight advance by one army, it is likely to capture a large number of the opposition's forces[1]).

The cube law indentifies a *'non-partisan bias'*, i.e. over time different parties are likely to take advantage of it in turning slender majorities (or worse) in terms of votes into workable majorities in a legislature. From this point of view it may be a desirable feature of an electoral system, in so far as it obviates the need for unwieldy coalitions between parties. There is, however, another side to this coin, which is usually termed *'minority-party bias'*.

There are numerous examples around the world of minority parties that find it very difficult to break into the mainstream of the political system (the British Liberals, the Irish Labour Party and the various European ecological parties, for example). The cause of this lies in the historical development of the political system, which is still represented in the existence of the electoral process as a spatial phenomenon. As Rokkan has argued, the original cleavages within society were spatial: this was expressed in unnumerable kingdoms, fiefdoms, palatinates and so on (1970; Taylor and

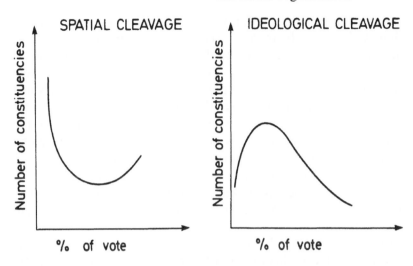

Figure 4.1   A stylistic representation of different types of political cleavage

The first graph illustrates what is termed here a spatial cleavage, i.e. a political party fights on a platform which has some locational component: this may be very general (e.g. a fishing or agricultural party) or very specific (e.g. a separatist party, cf. Quebec or Wales). Such a party picks up no votes (and probably has no candidates) in most seats; where it actually stands, it performs very well, however.

The second graph illustrates a more usual ideological cleavage (i.e. more usual in British politics). In this instance, parties fight on broad political issues (defence; spending; hanging) and thus pick up votes across the spectrum of constituencies. Only in very small number of constituencies are few votes recorded.

Johnston, 1979). As late as the turn of the century it was still possible to identify spatially based political parties; Taylor, for example, examines the brief career of the Irish Nationalists, who in 1910 polled only 1.9 per cent of the British vote, but captured 12.2 per cent of the parliamentary seats at Westminster (1973). As Figure 4.1(a) indicates, this was because their basis of support was spatial, not ideological, and in consequence their concentration in a small number of constituencies meant that the votes they polled were all used to good effect (a similar phenomenon is to be noted in Northern Ireland today with the Unionist parties). Quite the opposite effect is indicated in Figure 4.1(b), in which the Liberal Party support in 1970 is illustrated. In that year the Liberals collected 8.6 per cent of the votes, but realised only 0.9 per cent of the seats at Westminster. As we can see, this is because the distribution is inverted: there is no spatial concentration of support, and consequently the Liberal vote is dissipated throughout the country.

This mismatch of a spatial electoral system and ideological parties only becomes a problem if there also exists, as in Britain, a plurality system rather than some form of proportional representation (see Chapter 7). A problem does exist, moreover, in two senses. First, there is inequity with regard to the Liberals (and other ideological minorities). Secondly, the system at present encourages spatially based parties who can reap the benefits of appealing to distinct electoral concentrations such as the Welsh, the Scots and the Ulster Irish. In itself this is not an issue, although one may question the validity of major concessions being wrung from governments by relatively small spatial minorities with only limited numbers of voters behind them (this is a theme examined further below). A more worrying aspect is the way in which other types of parties may claw their way to prominence, by building upon a small concentration of support.

Before taking this theme any further, it is worth making a brief detour in order to illustrate how this process of establishing a party may operate. This has been well covered by Gudgin and Taylor (1974), in their analysis of the rise of the Labour Party, beginning with the latter's intervention in the General Election of 1910. At that time Labour support was both ideological *and* spatial, i.e. the party fared well in constituencies where manual employment was concentrated, such as South Wales, Yorkshire, Lancashire and the North East. This density of support permitted Labour to perform well, to the extent of winning 6 per cent of the seats with only 8 per cent of the votes. The importance of a spatial base was emphasized by the results of the 1918 election, in which Labour 'broke out' and contested 359 seats. The party's vote rose to 22 per cent of the total, but the dispersion of the Labour vote outside the core constituencies was emphasized by the poor rate of success: only 8.5 per cent (sixty seats). It was not until 1929 that Labour consolidated its support outside the coalfields and the industrial towns, and overcame this negative bias.

The suggestion that racialist parties could emulate this process may appear initially to be far-fetched, but there is some evidence that this is feasible. Recent research by Peake suggests that the National Front has in fact already attempted to speed up this cycle of growth, for in 1970 they stood in only 10 constituencies, all of which displayed a particular inner-city profile, the most notable aspect of which was a relatively high concentration of ethnic minorities. However, by October 1974 they had already attempted to break out, by standing in 90 constituencies. (In 1979 this figure had enlarged again to over 300 constituencies (Peake, 1980)).

As far as the National Front is concerned, the obvious difference between its growth and that of Labour is that the former has not attracted any substantial voter support. None the less, there exists evidence that racial politics can exist alongside the more familiar cleavages.[2] Taylor and Johnston for example discuss the ephemeral success of Enoch Powell in the West Midlands in attracting a large amount of political support on what was perceived to be a racial platform, although this, of course, was in an area with a relatively high concentration of coloured people (1979, pp. 294–300). They conclude:

> there is a clear inference . . . that the more volatile of the British constituencies in the last twenty years, in terms of swings of opinion from one main party to another, have included many with above-average percentages of immigrants among their residents. (1979, p. 300)

To date, the National Front and other small right-wing parties have adopted a stance that is too extreme to capture widespread electoral support in these constituencies, but this would not preclude a different party from exploiting these spatial concentrations at some future date, and perhaps then following through the growth cycle once more, with greater success. Certainly, the success of the Provisional IRA in a by-election in 1981 underlines the fact that extreme political platforms can be electorally successful as long as the spatial basis of support is sufficiently concentrated.

To conclude this section, mention must also be made of a corollary of spatial concentrations of support, namely that once a party is established, blocs of votes are counter-productive. Gudgin and Taylor write 'now that Labour is a major party, this heavy concentration of the vote is a liability. All these strongholds represent large numbers of excess votes which would be much more useful to Labour if they were distributed elsewhere' (1974, p. 69). Nor is this a minor problem: estimates have suggested that up to 2 per cent of the overall support (i.e. 500,000 votes) is 'lost' as far as Labour is concerned; this phenomenon is also to be noted in Australia and New Zealand.

To summarize this section we can isolate two themes. First, and to underline the overall theme of 'space and society', it is worth reiterating that the political arena may not simply be an ideological struggle, but a series of spatial contests as well. This should not be overplayed, in so far as the isomorphism once visible between party and place has disappeared (cf. the Irish Nationalists). None the less, the very existence of constituencies introduces a spatial element, permits the possibility of certain types of localized support (cf. the National Front or the Provisional IRA), and introduces types of electoral bias. This in turn relates back to the point emphasized within this section, namely that individuals can experience location-specific deprivation. This occurs in several ways, but notably because minority parties suffer from non-partisan bias, and majority parties waste a number of votes. The individual can, of course, dispose of his vote in any way he chooses, but in some constituencies a Liberal or Labour vote is a unit of devalued currency:[3]

for the 1955 election, Labour suffered a negative bias of approximately 5 per cent, which was made up of about 1 per cent due to their winning larger constituencies, 2 per cent due to their particular distribution of support and a final 2 per cent simply because they were the losing party. (Taylor and Gudgin, 1976a, p. 19)

Partisan bias and malapportionment

The above quote, besides quantitatively defining the extent of bias in a particular election, introduces the notion of electoral abuse, or the deliberate manipulation of the spatial basis of an election in order to influence the result. In this case it is suggested that Labour made poor use of votes because they prospered in larger constituencies than those being won by their opponents; in other words, votes were being wasted.

This may not, at first sight, appear to be an example of electoral abuse; after all, it would be very difficult to ensure that all constituencies were exactly the same size, due to the constant shifts of population that occur both within cities and between regions. However, there is good evidence that on the one hand the normal degree of variation transcends the accidental, and that on the other the organizational steps needed to rectify these situations are subject to political decision-making.

First, as far as the size of constituencies is concerned, it is easy to show that the spatial dynamics of population change have on many occasions produced a mismatch between the location of seats and the concentrations of votes. The 'rotten boroughs' of the early nineteenth century, where the MPs occasionally outnumbered the voters, were the most flagrant example of this imbalance; interestingly, their ultimate disappearance in 1832 marks a political watershed between the 'agricultural' bases of power, typical of the eighteenth century, and a recognition of the industrialization and urbanization that was to follow throughout the nineteenth. Even today, population growth and differential patterns of migration result in major anomalies between constituencies, with inner-city seats in particular displaying falling electoral registers. Several inner-urban constituencies have fewer than 35,000 voters, whilst expanding areas may have in excess of 80,000 or 90,000 constituents. The net result is once more that votes are devalued in certain instances; Taylor and Johnston use the example of the constituencies of Halesowen and Stourbridge (72,596) and Ross and Cromarty (20,402). In the former it required in excess of 34,000 Conservatives to carry the seat; in the latter only 8000. Simple arithmetic suggests that a vote in Ross and Cromarty is four times more important in terms of influencing the ultimate electoral outcome than in Halesowen (1979, p. 453).

These types of anomaly are supposed to be monitored and then removed by redistricting exercises, which redraw constituency boundaries to take into account population change. In the United States this process has always carried with it political overtones, and the term 'gerrymander' enshrines the name of Governor Gerry of Massachusetts, who was adept at the production of spatial solutions that rendered results favourable to his political needs. (Gerrymandering is normally achieved by splitting up the opposition's concentrations of support, and swamping each block in a larger group of partisan areas, although a full explanation of 'stacking', 'cracking' and 'packing' may be found in Johnston, 1979b, pp. 172–3).

In Britain the process is commonly assumed to be apolitical; none the less, partisan solutions may come about in two ways. The first involves a process of benign neglect, or 'silent' gerrymandering: in other words, if redistricting might be harmful to the incumbent party, no action is taken. Any doubts that the recommendations made by those responsible for redistricting (the boundary commissioners) are accepted either without demur or interference, can be dispelled by consulting Richard Crossman's memoirs of his career as a government minister:

for some eight years now the Local Government Boundary Commission has been at work . . . . Some of their most ambitious work has been done in the West Midlands conurbation and in Tyneside. All these proposals are now coming up to me, because in the last period before the election my predecessor, Keith Joseph, found an excuse for postponing any decisions.

Joseph's prevarications are not explained; Crossman does, however, provide some detail about the parliamentary implications of changing local government boundaries; he was told that

if the reform of the boundaries of Leicester went through, at least two of the Labour seats would be in danger . . . I also discovered that in Coventry there were risks involved, *but that I could by a minor amendment make practically sure that Coventry remains our way* (my emphasis). (1979, p. 45)

These remarks show conclusively that this supposedly neutral process of boundary realignment has in fact significant political overtones.[4]

The second way in which partisan solutions come about is less obvious, and relates to the methods used by the boundary commissioners themselves; (in this context, the emphasis is more upon the work of the *parliamentary* commissioners, not those responsible for local-authority boundaries). Their aim is the creation of constituencies that have (in England) a population within a range about a target of 57,122, although there is a necessity to observe administrative constraints. In effect, parliamentary constituencies should not cross county or district boundaries. Most importantly, the Commission is politically neutral, and not concerned with the political implications of its decisions.

In practice the Commission's job involves clustering wards together in order to create constituencies. Taylor and Gudgin, for example, discuss the examples of Newcastle and Sunderland, in which there are respectively 20 and 18 wards, from which four and two seats are created. With this number of units, there are a large number of possible solutions: 87 in Sunderland, 337 in Newcastle (1976b, pp. 50–1). Naturally, different configurations yield different results; as Figure 4.2 indicates for Newcastle upon Tyne, an overall political split (in terms of votes) of 58:42 in favour of Labour can be manipulated to yield results that range from 2–2 (Labour and Conservatives each winning two parliamentary seats) through to 4–0 (Labour takes all).

The Boundary Commission does not use computers, and is unlikely to investigate all the possible alternatives. This means that it

---

Figure 4.2   Possible methods of creating constituencies in Newcastle upon Tyne (source: Gudgin and Taylor 1976a, b)

There are twenty electoral wards in the city, and these are combined to create four parliamentary constituencies; there exist 337 means of achieving this. On the right the diagram shows two of the more 'extreme' solutions: the top one shows the best possible result for the Conservatives, i.e. winning two seats, each of which picks out the high-status areas to the north of the city, well away from the Labour seats in the vicinity of the River Tyne. A different partition, in which each seat abuts the river, produces a 4–0 win for Labour.

This histogram bottom-right indicates the 337 possible solutions, expressed in terms of F-ratios (see equation p. 86). The higher the F-ratio, the greater the dissimilarity between the seats: the Conservatives do well (2–2) when the seats are sharply differentiated, whereas the Labour party scores successes when the four seats are very similar, and Conservative support is just overcome in each constituency. The bulk of the solutions, it can be seen, yield a 3–1 solution: the actual result in the 1979 election.

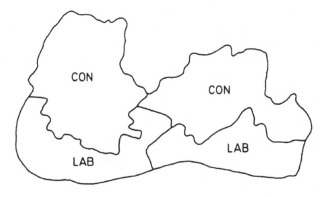

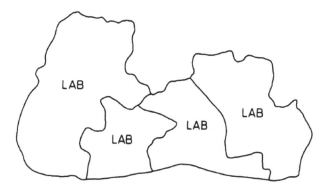

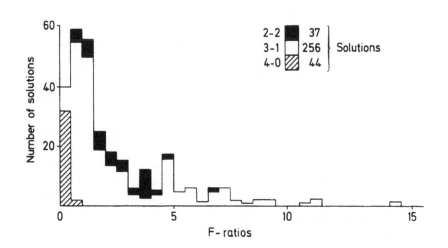

will opt for any 'reasonable' solution, i.e. one that fulfils its legal obligations concerning size and boundaries. As we can see from the histogram, this means that there is a three in four chance that it will hit upon a 3–1 solution, and even a one in seven chance that a 4–0 solution could arise. The F-ratio simply measures the ratio of between-constituency variance to within-constituency variance:

$$F = \frac{\text{Between-constituency variance of Labour vote}}{\text{Within-constituency variance of Labour vote}}$$

Consequently, high F-ratio solutions are those in which the boundaries pick out pockets of Labour support, and isolate Conservative voters in one seat; the corollary of this is a situation in which all four seats are very similar (low between constituency variance) and there is high within-constituency variance; in other words, the Tories just lose all four seats. As we have seen, the configurations that yield this result are not in a majority, and it should come as no surprise to learn that the 1979 election result was a 3–1 win for Labour.

So far we have simply seen the commissioners to be prisoners of the law of averages; from this we need, however, to progress to the statement that their activities lead to a form of partisan bias that automatically favours a particular political party. This is counter-intuitive; we do not expect neutral bodies to produce the same results as gerrymanderers; none the less 'although the decision-making procedure is different, the political result is similar. This paradox is a function of the fact that the aggregating procedure involved in all districting tends to favour the majority party unless specifically designed not to do so' (Taylor and Gudgin, 1976b, p. 55). This, simply because it is usually very difficult to partition areas to give proportional rewards (it is not possible to translate a 58:42 distribution of votes into an equitable balance of four seats, for example), 'we can view the work of the Boundary Commissioners in Britain as going around the country drawing lines that usually favour local majorities, such as Labour supporters in industrial towns, and Conservative supporters in rural or suburban counties' (Taylor and Gudgin, 1976b, p. 55).

Seen more simply, small pockets of minorities are liable to be swamped in any districting exercise, biased or not. We can extend our argument, therefore, to suggest that the manipulation of the spatial base of elections produces class-specific deprivation. This may be an overt process of political decision-making (gerry-mandering) or the unintentional outcome of a supposedly neutral action. Either way, particular political groups, usually those in a minority within a particular district, will not be permitted to put their votes to the fullest use.

Political support and financial rewards

The preceding argument has posited a close relationship between the political process and the spatial bases of support. The extent to which it makes any sense to think of British politics as a whole as a spatial issue, rather than an ideological one, will be summarized below. None the less, we can throw additional light on the spatial nature of support by inverting the argument, and examining the ways in which those with political power attempt to reward their supporters, and attract future voters; in the American literature, this process is graphically compared with the division of the contents of the 'pork barrel.'

A successful government can reward its supporters in two main ways. First, it can pass legislation of a general nature, designed to reward or appease voters; such measures could be reformist (the changing of divorce laws, for example) or economic (the alteration of the personal tax structure, for instance). Secondly, and from our point of view more interestingly, it can make decisions that have explicit spatial consequences:

governments must take both positive and negative decisions, especially in periods of retrenchment. Positive policies involve the allocation of public goods, such as investment grants in particular places, whereas negative policies involve the removal of public money from an area, as with the closure of a defence establishment. The former are vote-winners, and will be directed at marginal seats in some cases; the latter are vote-losers, and will, if at all possible, be directed away from marginal seats. (Johnston, 1976, p. 191)

This situation is succinctly described by Johnston as 'the geography of Downs' (alluding to the work of Anthony Downs), which envisages political parties not only cutting their ideological cloth in order to capture the centre ground of opinion, but also identifying the particular constituencies (the marginals) by which this may be achieved (1979b, Chapter 6).

The examples chosen by Johnston are varied, although several are, in his words, 'quasi-anecdotal'. The cases he cites vary from a by-election in Hull (at which time the long-awaited Humber Bridge was approved as a planning project), through to the longer-term policy of appeasing devolutionary parties in Scotland and Wales with political support in the House of Commons. A more rigorous analysis, which examines the closure of teacher-training colleges, indicates that 'the decisions on which colleges to close were clearly taken with electoral considerations in mind, although significance tests only partly support this interpretation' (Johnston, 1979b, p. 119).

In this sense, therefore, political ideologies may be subsumed into a more pragmatic stance which takes account of spatial variations in support, and any local needs which may be 'bought-off'. Nor is this simply manifest as a series of isolated rewards; it is far more usual to be able to identify long-term policies which favour particular areal concentrations of support. In the United States 'pork barrel politics' constitute an accepted tradition in which representatives strive to obtain rewards for their constituencies, either in the guise of government spending (on for example defence contracts), or perhaps in the form of avoiding some externality, such as a nuclear-weapons establishment (see, for example, Chapter 6). In Britain Labour governments have more strongly favoured the Assisted Areas (which contain many Labour voters) and the large metropolitan centres, which also contain Labour seats, and which have been favoured by the weighting of the Rate Support Grant[5] (see Chapter 7).

As always, some balance is required with this type of interpretation. Discussing the initiative directed to the inner city by the Labour administration in 1967–8, Johnston remarks, 'the latter areas provide some of Labour's safest seats: but these are among the country's smallest, and their populations have been declining. If the decline were not halted, therefore, safe seats could disappear', 1979b, p. 116). This is clearly factually correct, but oversimplifies the position. Other interpretations of this history have emphasized the long-term self-interest that prompted the state to act; the Community Development Project report *Gilding the Ghetto*, for example, points out that the urban initiatives were conceived by Callaghan while he was Home Secretary, and funded thereafter from the Home Office, the government department responsible for – in crude terms – law and order, not social services or economic planning (1977). From this it infers that cash was being spent in order to defuse racial, and potentially political, unrest. Whatever the interpretation, the relationship between policy making and a recognition of the spatial nature of both political problems and potential support is not in doubt. Nor should it be overlooked that those types of explanation go some way towards accounting for variations in public provision between areas with similar needs, as noted in Chapter 2.

The neighbourhood effect in 'Thatcherland'

To conclude this chapter we may usefully examine in greater detail the question posed above. Namely, to what extent is there some tension between the spatial basis of elections, and the existence of ideological political parties? We have seen that the plurality system produces non-partisan bias, and that the need to change and redistrict constituencies provides scope for 'partisan cartography'. Furthermore, the existence of recognizable blocs of support produces an additional secular dimension into decision-making, with governments avoiding measures that may be desirable, but are politically inexpedient *vis-à-vis* a few marginal seats. Does this add up to a prognosis that modern parties are shackled by a constituency division of support, and that aspatial elections would remove numerous iniquities? An answer to this question has three parts, relating to the 'pork barrel' and the interrelated topics of neighbourhood effects and spatial cleavages; all these terms will be explored below.

First, it should be pointed out that even if constituencies were to disappear, spatial lobbies would not. Space is a highly predictable commodity, with the 'laws' of real estate acting as an efficient arbiter of land use. There is thus a repetitive pattern of, say, high-status suburb and low-status inner city throughout the western world, and in each part of each city there exist conforming patterns of political outlook on issues such as transport, housing and planning. This is particularly true in the United States, where this core/suburb split is particularly marked. One of the more obvious results of this creation of legally bounded spaces by and for different communities is that some cities, divorced from wealthy suburbs, have low rate or tax incomes, which has led in large measure to the present urban fiscal crisis (Bennett, 1980). On a larger scale we might also identify different types of predominant economic activity; consequently, farming zones, fishing ports or regions with particular economic problems (relating perhaps to structural change) will also function as spatial lobbies on issues as diverse as manufacturing imports or membership of the EEC. As a result, a removal of constituencies would not remove the 'pork barrel' as well.

The second point relates to the question of political thought and voting behaviour. We have already confronted the suggestion that political outlooks were originally moulded by locational interests, but that these were superseded by political allegiances of the type familiar in Britain today. Throughout most of this century the main cleavage has been dictated by attitudes towards the redistribution of wealth, although we have, of course, also noted Dunleavy's views on the amelioration of this effect as a result of consumption issues (Dunleavy, 1980). One logical outcome of this development is a change in the rationale of support from simple locational factors to what usually reduce to socio-economic factors, which can be collapsed even further to 'self-interest'.

This is, however, not the only determinant of support for a political party: much research has identified the existence of contextual factors, the most important of which is the so-called *neighbourhood effect*, which causes individuals to be 'infected' with the political ideas of their neighbours:

Some high-status London suburbs contain council-housing estates and most mining areas will have pockets of middle-class housing. Each of these social areas is a separate social milieu, with a dominant political ethos. But none is entirely independent of those surrounding it. They may be grouped together to form catchment areas for schools, churches and various social organisations, within which are brought together people from different backgrounds. And so each little part of the country has its own dominant social ethos and political orientation. (Taylor and Johnston, 1979, p. 234)

The neighbourhood effect has been examined in several contexts, and at various spatial scales. As far as parliamentary constituencies are concerned, Taylor and Johnston show that there are two sorts of process at work. As the proportion of a particular social group increases (e.g. the percentage working class), two things happen. First, voters who might be expected to vote Conservative instead support Labour. Secondly, the proportion of the working-class group that opts for Labour also increases: in 1969, for example, constituencies with low concentrations of working-class voters saw only 1 in 2 of the latter voting Labour; however, in seats where there was a majority from the working class, this proportion increased to 2 out of 3 (1979, pp. 255–9).

A different manifestation of the neighbourhood effect depends not upon political discussions within areas, but rather a simpler tribal loyalty; this has been termed the 'friends-and-neighbours' effect, and results in a candidate attracting votes simply because he is a 'local boy'. The most famous study of this involves the analysis, by regression, of the vote for Senator McCarthy in Winsconsin, using a simple measure of rurality as the independent variable. The latter provides a good explanation of his support, except for heavy under-prediction in the east of the state, and it comes as no surprise to find that this was the home county of the Senator (McCarty on McCarthy, 1952). Clearly, this is a factor that is difficult to predict from election to election; it is noteworthy, however, that Cox suggests that in excess of 10 per cent of the electoral vote may be attributed to contextual effects (O'Loughlin, 1981).

Although these analyses are not concerned with next-door-neighbours, they have, none the less, tended to concentrate upon individual constituencies. Is it possible to build upon these remarks to isolate the existence of broader spatial cleavages? The evidence is contradictory. If, for example, we examine the results of a national referendum like that dealing with EEC membership in 1975, we find a vote that mirrors 'normal' political cleavages, rather than a core-periphery pattern in which those far from central authority oppose another, yet-more-distant layer of bureaucracy (Kirby and Taylor, 1976).

A different conclusion is reached by Webber, who identifies 'the new geography of party allegiance' (1978). By examining voting swings between 1955 and 1974, he is able to identify six types of constituency, which although not spatially contiguous are typified by consistent electoral results. These are summarized in Table 4.3.

Webber concludes that

this new material demonstrates that though Britain will continue to divide politically on a class basis, there are many other dimensions which need to

Table 4.3 Groups of constituency, based on swings to Labour

Description	% Swing to Labour (1955–74)	Examples
(1) Suburbs and service centres	+1.0	Manchester; Southampton; Chipping Barnet; St Albans
(2) Rural areas and seaside resorts	−2.0	Oswestry; Sudbury; North Angus; Morecambe
(3) Growth areas	−2.9	Bedford; Derbyshire S.E.; Grantham; Meriden; Horsham
(4) Stable industrial areas	−0.6	Westhoughton; Barnsley; Dudley E; Stockton; Aberdare; Wakefield; Stretford; Bolton E.
(5) Areas of much council housing	+5.3	Jarrow; Nottingham W.; West Lothian; Edinburgh E.; Glasgow, Maryhill
(6) Metropolitan inner areas	+4.7	Lewisham; Nottingham E.; Camden; St Pancras N.

Source: after Webber (1978).

be considered. These include urban/rural, growth/decline, family/single people, public/private housing, service/industrial employment. Each of these divides, except the last, seems to be becoming more influential than it used to be. (1978, p. 683)

The conclusions were borne out by the results of the 1979 election. In two complementary papers which discuss the hypo-thesized emergence of 'Thatcherland', an apparently solid body of Conservative support in the south of England, Taylor and Johnston identify similar factors at work; in fact Taylor (1979) identifies *four* spatial areas of voting allegiance:

(1) *the exurban south*: a region with the highest swings to the Conservatives;

(2) *outer metropolitan England*: areas outside the conurbations;

(3) *inner metropolitan England*: the conurbations of the Midlands and the North;

(4) *old industrial hearths*: the coalfields and the traditional Labour areas.[6]

Taylor's paper provides a useful conclusion to this section, as he assesses the likely implications of a growth in spatial cleavages of this type; (see, however, 'Postscript', below):

the increasing segregation of support patterns will affect the whole electoral system. The responsiveness of the system to the voters depends on the relationship between the patterns of support and constituency boundaries. Increasing separation of partisan voters will lead to more safe Labour seats in urban areas and more safe Conservative seats outside the urban areas, producing a tendency for marginal seats to decline. Of course, less marginal seats means less seats changing hands at election time, so that elections become much more stable affairs. It is no overstatement to suggest that spatial demographic change is undermining the operation of our elective-democratic system through its effect on the geography of representation. (1979, p. 293)

By way of summary, therefore, we can say that the spatial factor in the electoral system is far from straightforward. Although ideo-logical parties exist here in Britain, it is clear that their bases of support can still be easily identified in a spatial context, and even that regional patterns of support may be emerging. In one sense this may reflect a return to the 'normal', locational pattern of political rivalry; in another, it may impose a rigidity upon political life that some, like Taylor, regard with disfavour.

Postscript

Clearly, the remarks made above are based upon interpretations of the May 1979 election results. Politics, within Britain at least, have, however, evolved since that date; thus whilst at the time of writing it is still possible to make firm predictions about, say, the spatial distribution of support for the American Equal Rights Amendment, it is suddenly less possible to predict the results of the next general election in the UK.

The cause of this potential confusion lies in the creation of a new political entity, namely the Social Democratic–Liberal Alliance, which has evolved from those leaving the Labour Party and the erstwhile Liberal Party. The Social Democratic Party (SDP) has existed only since July 1981, and the Alliance has been in operation only since September of that year; consequently it is difficult to make detailed inferences. None the less, a study of the *local* (ward) seats won by the constituent parts of the Alliance provides some interesting information. Between July and December 1981 the Liberals won 61 seats, with a sizeable proportion being taken from the Conservatives; this builds on past successes, and is normal during the mid-term of an unpopular government. The SDP, however, has no political past, and its 28 wins are important for that reason alone. Interestingly, however, a large proportion (60 per cent) of the SDP's gains have been from the Labour Party; in other words, the SDP has begun its self-appointed task of replacing the Labour Party as the main 'socialist' party in the country. Of even greater importance is the additional fact that this process has not depended upon any particular spatial concentration of support for the party, as Table 4.4 reveals.

From Table 4.4 we can see that support for the SDP appears to be national, with seats throughout England and Wales being represented. Clearly, it is dangerous to submit the recent past to detailed scrutiny; none the less there is little evidence that the new party is based upon a solid spatial core of support, as occurred with the inception of the Labour Party. Expectations that the SDP might be based solely in what Donnison and Soto characterize as 'New Britain' (1980), are dispelled by local wins in metropolitan wards like those in Islington and Sefton, or in coalfield constituencies like Sedgefield or Bedwellty.

If it is dangerous to interpret the past, it is of course folly to predict future performances. In Table 4.4 the local successes have been related to the parliamentary seats in which the wards are located, and some simple indicators from the 1971 Census dis-

Table 4.4 Chronological list of electoral gains (in district, borough or county council elections) by the Social Democratic Party in Britain, June–December 1981; data for related parliamentary constituencies* (1971)

SDP gain (district or borough location)	Related parliamentary constituency	1971 Census Indicators		
		% Owner occupation	% Car ownership	% New common-wealth origins
Vale of White Horse	Abingdon	52.6	55.1	1.3
Sedgefield	Durham	38.3	39.2	0.3
Newcastle upon Tyne	Newcastle (4 seats	–	–	–
Hemel Hempstead	Hemel Hempstead	40.8	52.3	1.1
Lambeth	Lambeth (2 seats)	–	–	–
West Oxfordshire	Mid-Oxon	55.6	58.7	1.5
Islington	Islington (2 seats)	–	–	–
Leeds	Leeds (5 seats)	–	–	–
Newbury	Newbury	56.0	53.7	1.1
Alyn and Deeside	E. Flint	56.9	51.4	0.3
Tendring	Harwich	72.2	43.8	0.7
Halton	Runcorn	61.6	50.1	0.4
Sefton	Crosby**	68.5	48.3	0.5
Merton	Merton (2 seats)	–	–	–
Monmouth	Monmouth	50.6	53.3	0.4
Harrow	Harrow	–	–	–
St Pancras	Holborn	–	–	–
Southwark	Southwark	–	–	–
Mid-Sussex	Mid-Sussex	66.4	53.4	1.4
Hemel Hempstead	Hemel Hempstead	40.8	52.3	1.1
Selby	Barkston Ash	61.4	50.1	0.4
Lancaster	Lancaster	60.8	42.6	1.0
Tendring	Harwich	72.2	43.8	0.7
Bridgenorth	Ludlow	49.9	50.1	0.4
Adur	Shoreham	66.7	47.6	1.0
Islwyn	Bedwellty	42.6	41.6	0.2
Cleveland	Cleveland & Whitby	58.9	45.2	0.4
Canterbury	Canterbury	61.9	43.5	1.1

*Where a local seat cannot be related directly to one parliamentary constituency, no data are presented.
**SDP gain, November, 1981.

played. In keeping with the remarks already made, we can see that *if* the SDP fights these seats (where it already possesses a party machine), it will be contesting a broad range of constituencies, in which wealth (car ownership), consumption (public housing) and ethnicity all vary widely in importance. A majority of the seats is Conservative, but interestingly at least ten of these seats could be captured by a coalition of existing Labour and Liberal votes.

No comments are made here concerning the likely results of by-elections and general elections in coming years. What seems clear, however, is that the broad spatial perspective discussed in relation to 1979 may already be dated. The speed with which the SDP has established itself breaks many of the unwritten rules concerning the growth of parties in plurality systems; the reasons why this has occurred are, however, relatively simple. In many senses the SDP's growth began some decades ago; it is, in other words, not really a totally new party, and it has thus been able to build upon established Labour Party support, and in some instances even upon complete local party machines.

Explanations for the 'sudden' success of the SDP are varied, but most observers emphasize the inability of the Labour Party to locate itself correctly along the ideological continuum (see, for example, Johnston, 1979b, pp. 81–7). Himmelweit *et al.*, for instance, note that the grounding of Labour's policies in class or production issues is largely irrelevant in an era when consumption questions cut directly across class lines (Dunleavy, 1982; Himmelweit *et al.*, 1981). Given this interpretation, it is the failure of the Labour Party rather than the SDP's intrinsic attractions which accounts for this electoral change, and the enormous changes to the electoral map which are likely to result.

Conclusions

This chapter has attempted to untangle two themes. The first is that the political process within a society such as ours is explicitly spatial, in so far as it takes place within a constituency system at different geographical scales. This has various implications, but the most obvious one is that there exists a tension between the spatial nature of political organization and the ideological basis of much political activity and thought. At the simplest level, policy making can be 'secularized' to take account of the 'pork-barrel'; at the other extreme, we find that some political parties can exploit small spatial concentrations of support in order to further their electoral prospects. In itself, this can be seen as part of the plurality of a parliamentary system, in which shades of opinion can be accommodated. However, it is also to be considered in the context of the second theme explored here, which is that the spatial organization of the political process must lead to inequalities between voters, or as we have used the term above, deprivation. Thus any electoral success of, say, the National Front, must be balanced against the relative demise of the Liberal Party in the past, which is to be explained in terms of the inbuilt biases within the electoral system. As we have seen, there exist various types of bias which affect both locations and certain groups of voters. Residence within a particular constituency can produce virtual disenfranchisement for some voters, either because of their preferred support or more simply due to the different forms of malapportionment that perpetuate. As we shall see in Chapter 7, quite major changes in the electoral system are required to overcome these inequalities.

Notes

1 Rather more sophisticated proofs of the operation of the cube law are available in Johnston (1979b, pp. 58–63) and Taylor and Johnston (1979, pp. 392–6).
2 The obvious source of data in this instance is the USA. O'Loughlin and Berg (1977) show, for example, how concentrations of black voters tend to bring out bloc votes from whites in mayoral elections; Cox in fact argues that there exists an *inverse* relationship between black concentrations and electoral success (1973).
3 There are, of course, some constituencies in which the Conservatives, or minority parties such as the Unionists, suffer from wasted votes, but these are quantitatively less of a problem.
4 A fact not overlooked by Crossman; he writes 'I find myself as Minister of Housing a powerful politician in my own right'. Later, he is reminded by the Prime Minister 'to make quite sure that I was doing my job as a politician on the local boundary decisions, that no adjustment was politically disadvantageous' (1979, pp. 45 and 89).
5 This process of rewarding voters may also work in reverse: Labour councillors in Sheffield have suggested that wards in the city that had voted Conservative should be made to bear the brunt of the 1979 local-authority spending cuts, initiated by the new Conservative government.
6 Johnston's paper is of interest in that it examines the neighbourhood effect that led to the different patterns of voting behaviour throughout the country (1979c).
7 The reader should not, of course, overlook that success at the ward scale does not guarantee automatic success at the parliamentary scale; such an assumption would be an ecological fallacy. None the less, the 1981 Crosby by-election win for the SDP indicates that parliamentary success is likely to build on ward victories.

References

Bennett, R. J. (1980) *The Geography of Public Finance*, London, Methuen.
Community Development Project, (1977) *Gilding the Ghetto*, London, National Community Development Project.

Cox, K. R. (1973) *Conflict, Power and Politics in the City*, New York, McGraw-Hill.

Crossman, R. (1979) Howard, A. (ed.) *The Crossman Diaries*, London, Methuen.

Donnison, D. and Soto, P. (1980) *The Good City*, London, Heinemann.

Dunleavy, P. (1980) *Urban Political Analysis*, London, Macmillan.

Dunleavy, P. (1982) 'Response to Hooper', *Political Geography Quarterly*, 1(2).

Gudgin, G. and Taylor, P. J. (1974) 'Electoral bias and the distribution of party voters', Institute of British Geographers, *Transactions*, 63, 53–74.

Himmelweit, H., Humphreys, P., Jaeger, M. and Katz, M. (1981) *How Voters Decide*, London, Academic Press

Johnston, R. J. (1976) 'Resources allocation and political campaigns', *Policy and Politics*, 5, 181–99.

Johnston, R. J. (1979b) *Political, Electoral and Spatial Systems*, Oxford, University Press.

Johnston, R. J. (1979c) 'Regional variations in the 1979 General Election results for England', *Area*, 11(4), 294–7.

Kirby, A. M. and Taylor, P. J. (1976) 'A geographical analysis of the voting pattern in the EEC Referendum, June 5, 1975', *Regional Studies*, 10, 183–91.

O'Loughlin, J. and Berg, D. A. (1977) 'The election of black mayors', Association of American Geographers, *Annals*, 67, 223–38.

O'Loughlin, J. (1981) 'The neighbourhood effect in urban surfaces', in Burnett, A. and Taylor, P. J. (eds) *Political Studies from Spatial Perspectives*, Chichester, John Wiley.

McCarty, H. (1952) McCarty on McCarthy; see Thomas, E. N. (1968) in Berry, B. J. L. and Marble, D. (eds) *Spatial Analysis*, Englewood-Cliffs, Prentice Hall, 326–52.

Peake, L. (1980) 'An analysis of the electoral potential of the British National Front', mimeo, Department of Geography, University of Reading.

Rokkan, S. (1970) *Citizens, Elections, Parties*, New York, McKay.

Taylor, P. J. (1973) 'Some implications of the spatial organisation of elections', Institute of British Geographers, *Transactions*, 60, 121–36.

Taylor, P. J. (1979) 'The changing geography of representation in Britain', *Area*, 11(4), 289–93.

Taylor, P. J. and Gudgin, G. (1976a) 'The myth of non-partisan cartography: a study of electoral biases in the English Boundary Commission's Redistribution for 1955–70', *Urban Studies*, 13, 13–25.

Taylor, P. J. and Gudgin, G. (1976b) 'The statistical basis of decision making in electoral districting', *Environment and Planning*, A8, 45–58.

Taylor, P. J. and Johnston, R. J. (1979) *Geography of Elections*, Harmondsworth, Penguin Books.

Webber, R. (1978) 'The geography of party allegiance', *New Society*, 45, 682–3.

Part III

Conflicts

In Chapter 4 we briefly examined the formal electoral system. This, however, restricts the debate about the ways in which political arguments develop within society. One of the most interesting aspects of political conflict is that it occurs in numerous contexts: between nations, between regions, between tribes and religious groupings, and – of interest here – between particular locations. It is, of course, valid to point out that regions are also locations, but in such conflicts (as between Corsica and mainland France, or the Basque territory and Spain), the nature of political opposition becomes formalized. The emphasis in the following chapters is firmly upon the more ephemeral types of opposition that spring up, particularly as a result of locational conflicts (cf. Chapter 1, 'The location of externalities').

In Chapter 5 two broad locational issues are examined, and in both instances the emphasis is upon the way in which the planning system places major nuisances upon the landscape. The first example deals with the almost perpetual issue of a third international airport for London, whilst the second concentrates upon the deliberations that surrounded the Windscale Inquiry, dealing with proposals to expand the nuclear waste-processing plant. Chapter 6 takes the same theme – the creation of conflict over locational issues – but examines it at a different spatial scale. In this instance the impacts of road construction schemes and the urban renewal process are considered.

In each case an attempt is made to develop some theoretical understanding of the issues involved, a process that Castells has dominated by his examination of the creation of 'urban social movements' by urban renewal. Once more, therefore, the context is spatial, but the implications are far wider.

5 Spatial cleavages

We all live in Harrisburg
  (protest badge)

The growth of interest in the environment has been closely documented by Sandbach (1980), who shows how membership of groups
such as the Friends of the Earth and the Nature Conservancy
increased dramatically in the middle and late 1960s, and how
newspaper coverage reacted with concern to environmental issues
such as the Torrey Canyon debate. As he also goes on to argue,
however, an understanding of this development is not simply
gained. It is not enough to view the environmental lobby as some
post-hippy reaction to a technological world, a 'natural'
involvement by concerned and committed individuals in some
pluralistic political debate about how the world's resources should
be used. At the very least, it is well established that environmental
concern was very much – and remains – the fiefdom of a relatively
limited number of well-educated and generally high-income individuals (Sandbach, 1980, pp. 106–37). More importantly, it is clear
that environmentalism has established itself because there are
sound economic pressures upon advanced societies to conserve
many forms of resources, and consequently such concern has
become a part of the standard political ideology.

A further point, which is not explicitly faced by Sandbach, is the
fact that much environmental protest is fired by self-interest. This
has been termed 'conservative environmentalism', or more critically 'ecofascism' (Pepper, 1980, p. 181). The key here is that
protest, about noise or dirt or nuclear power, is spatially based, and
thus represents the response by particular groups of residents to a
very localized threat: a threat to their own territory. Not all spatially
based protest is so concerned with externalities; as Stephenson
points out, local environmental groups may simply be an organizational convenience (1980). However, the argument to be
developed here focuses firmly upon self-interest, and the way in
which threats to particular locations produce informal political
cleavages. These political responses are frequently short-lived, and

possess very limited objectives; they have little to do with what O'Riordan terms 'radical ecological activity'. The predominance, however, of such spatial protest goes some way to explaining the relative weakness of the ecological movement *vis-à-vis* issues like nuclear energy (Sandbach, 1980, p. 125).

This chapter examines the nuclear issue, but we begin with the best documented instance of a large planned externality, and the reactions to it.

The Third London Airport (TLA)

The story of the putative airport to relieve the traffic now using Heathrow and Gatwick begins in 1944 (before Heathrow itself came into major use), and is still not resolved as we enter the last two decades of the century. Indeed, a cynic might ask whether London will get its airport before fuel prices render it obsolete altogether. It is a story that has involved the biggest exercise of social scientific analysis ever staged in this country, and a public reaction as fervent as anything the nuclear debate can spark off: the cartoon in Figure 5.1 is one example.

The chronology of the TLA saga has already been well documented (Bromhead, 1973; Hall, 1971, 1974, 1979; McKie, 1973; Sealy, 1976). However, none of these studies has fully developed the history as an example of the way in which the planning process operates with regard to the siting of major, negative externalities on the landscape. Many of these studies are interested in the TLA *per se*: that is, in the locational issues that suggest one or other particular site. Here, I do not intend to make any particular judgement on the locational question; rather the intention is to consider the events, particularly those that relate to the Roskill period, in the light of a range of considerations. In order that these themes can be fully developed, however, it is necessary to establish the brief outline of events in the TLA deliberations.

A chronology of the TLA

Before the Second World War London depended upon Croydon for its air links, although it became obvious at the time of the capital's post-war regeneration that a new, greenfield site was necessary. As a consequence, Abercrombie's Greater London Plan promoted Heathrow to the forefront of London's airports, with Gatwick and Stansted being alternately promoted and rejected as sites for additional development and expansion. At the time of the 1954

Figure 5.1   Trog provides a widely held view of the effects of air traffic (and in particular Concorde); courtesty of *Punch*

White Paper the present situation had emerged; Heathrow was to take the bulk of traffic, with Gatwick absorbing peak-period flights. Stansted was demoted due to military needs, whilst Blackbushe, originally a contender, slipped from sight to cater for private flights and occasional rock festivals.

The necessity for a third airport emerged during the late fifties and early sixties, as air traffic continued to develop. In 1961 an inter-departmental committee of government officials, dominated by Ministry of Aviation as opposed to planning personnel, was set up to consider air-traffic requirements, and their advice in 1963 was that a much-needed TLA should be constructed at Stansted, in order to meet an over-stretching of Heathrow-Gatwick capacity, within a decade.

At this point the chronology enters the first of a series of endlessly repeating episodes. A governmental inquiry was set up in 1965; its report, published in 1967, was highly critical of Stansted's location and development in terms of noise and amenity criteria. Despite this criticism, the Labour Government opted for development at Stansted that same year. Public outcry was prompted, and on becoming President of the Board of Trade in August 1967, Anthony Crosland opted for another, more far-reaching inquiry, chaired by Mr Justice Roskill.

Roskill and after

Peter Hall has described the Roskill Commission as 'a heroic attempt to extend the field of rational, balanced socio-economic inquiry into a very difficult area of decision' (1971, p. 143). In contrast to the Stansted investigations of 1961, the members of the Commission were not tied to aviation, but instead had links with several aspects of planning and social science; amongst those on the Commission were Buchanan, a Professor of Transport; Walters, a Professor of Economics; Goldstein, a transport consultant, and Hunt, a planning inspector in what was later to become the Department of the Environment. Their aim was to undertake an enormous exercise in spatial analysis: to examine some seventy-eight possible sites, and to measure, with respect to several criteria, their usefulness as the TLA.

The 'long list' of seventy-eight sites, some of which could only function as a *national*, rather than a *London* airport (Ferrybridge in Yorkshire is an obvious example), was quickly whittled down to four distinct 'possibles', and these are indicated in Figure 5.2.

A choice was made between these four sites by the application of cost-benefit analysis (CBA) (Chadwick, 1971; Self, 1975). The CBA involved a whole list of quantifiable (and some argued, non-quantifiable) variables such as capital costs, travel times, destruction of agricultural land, loss of environment *et similia*. The aim was not to find the cheapest site in real terms, but to isolate the relative costs in each case; the results of the CBA are outlined in Table 5.1.

The table of the benefit costs reveals that in many respects, Maplin/Foulness is the 'best' site, i.e. the cheapest. It suffers, however, from high passenger-transport costs in relation to both London and to the country as a whole. Different values were placed upon these passenger-transport costs, yet even low rates cannot displace Cublington as the least-cost site overall. This was Roskill's conclusion, with one notable exception: Professor Buchanan's note

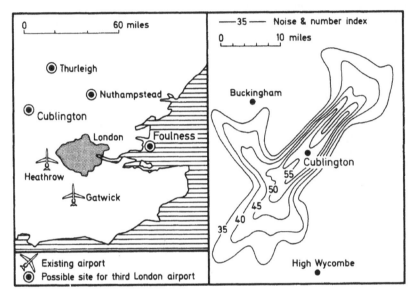

Figure 5.2   The location of the four 'Roskill' sites, and an example of predicted NNI contours around Cublington

The sites of Cublington (eventually chosen by the Roskill Commission), Thurleigh, Nuthampstead and Foulness/Maplin were given the greatest scrutiny, although several dozen possibilities were examined.

The NNI contours illustrate the degree of disturbance likely around any one of the possible sites.

of dissent, an 'evocation of the beauties of the Vale of Aylesbury' (Hall, 1971).

In many ways the minority report was the undoing of this enormous and costly (£1 million) exercise. Buchanan gave 'himself up as a hostage to all those whose approach to the question is visceral'. As McKie points out (1973), the release by the Commission of their decision attracted enormous environmental criticism before their methodology was ever analysed; in such a context Buchanan's prose was all the evidence that opposition

Table 5.1 Summary of the Roskill cost-benefit analyses: all costs in £m discounted to 1982, in excess of cheapest site

	Cublington		Foulness/ Maplin		Nuthamp- stead		Thurleigh	
	HTV*	LTV**	HTV	LTV	HTV	LTV	HTV	LTV
Airport construction	18		32		14		min	
Airport services	23	22	min		17	17	7	7
Passenger-user costs	min		207	167	41	35	39	22
Defence	29		min		5		61	
Off-site noise	13		min		62		5	
Agriculture	min		4		9		3	
Commerce and industry	min		2		1		2	
Extension of Luton	min		18		min		min	
Luton noise cost	min		11		min		min	
Meteorology	5		min		2		1	
Airspace movements	min		7	5	35	31	30	26
Freight-user costs	min		14		5		5	
Road capital	min		4		4		5	
Rail capital	3		26		12		min	
Public scientific establishments	1		min		21		27	
Private airfields	7		min		13		15	
Residential conditions (on-site)	11		min		8		6	
Public buildings (noise)	7		min		11		9	
Recreation (incl. noise)	13		min		7		7	
Aggregate expenditure above cheapest site	min		197	156	137	128	88	68

Source: adapted from Hall (1971) and GB Commission, Third London Airport (1971).
*HTV = high-travel time values
**LTV = low-travel time values (see text).

groups, including many MPs, required. In April 1971 Maplin, not Cublington, was chosen as the TLA.

Having overturned an "objective', 'scientific' analysis, it is not surprising to find that the Board of Trade took decisions which oscillated from site to site. As Peter Hall observes, political realities clearly had a role to play in this process, with a Labour opposition standing against a Foulness/Maplin site that was rapidly increasing in potential cost. Consequently, when Anthony Crosland returned in 1974 to Government, he rescinded the Maplin proposals and disbanded the Maplin Development Corporation (Hall, 1979).

The so-called Maplin Review introduced a new set of considerations into the whole air-transport issue: rising oil prices, slower growth of incomes and falling travel-demand projections. Within such a context a new set of possibilities also emerges. Simply, with the flexibility offered by the introduction of wide-bodied jets, Heathrow and Gatwick can continue to absorb greater numbers of passengers (but not greater numbers of flights), whilst Luton and Stansted could be increased in terms of capacity: (this is the situation broadly outlined in the 1978 White Paper *Airports Policy*). However, there is no guarantee that this will be the last word on the topic; at the time of writing another public inquiry at Stansted has commenced.

The implications of the TLA

We can see the information that has been briefly outlined above in a series of different ways. Hall has considered the history of the TLA as a 'planning disaster', and has examined the role of the actors involved (1979). Sealy on the other hand has documented the issue within the narrower, more technical confines of air-transport strategy within the UK (1976). As I have already suggested, the information available to us is useful in a very different context, namely the impact of planning decisions in a spatial domain. Essentially, the relationships to be developed are outlined by Batty (1979), and we can usefully use this diagram as a starting-point (Figure 5.3). In what, it must be stressed, is a stylistic view of the planning of reality, we may note three connected influences which can be assumed to be acting upon reality: the planning system itself, the political system and the community (in this instance the Third London Airport constitutes our reality). Let us begin with the planning system.

The planning system

Although it can be argued that the British planning system as a whole lacks any overall social goals or aims, and that in consequence certain incorrect (or at least suboptimal) decisions have been made (Kirby, 1981a), we must come to terms with the possibility that in some instances the planning system itself functions rather like some of the activities that it attempts to manipulate: that

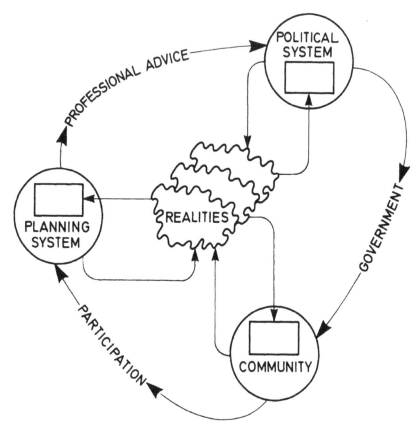

Figure 5.3   Components within the planning process: a systems view (from Batty, 1979)

This stylistic representation is used to indicate possible relationships between entities such as the planning profession, politicians and residents. The diagram is thus a visual device rather than a formal statement of the relationships that exist between these groups, and no reference back to a systems view (structural–functionalist view) of society is intended.

is, it is responsible for locating on the landscape major negative externalities. In the present example we are discussing with some detachment the location of an airport, which can in two ways have a major impact upon communities scattered over a wide area.

First, and perhaps most obviously, there is the noise issue. In the case of aircraft the intrusion of engine noise is so common, that standard measures of sound levels have been devised (see Sealy, 1976, p. 20). Although we cannot assume that all households are equally affected, the 35 NNI (Noise and Number Index) is frequently taken as a watershed between quiet and noisy areas around airports. As Figure 5.2 indicates, the 35 NNI boundary extends for some miles around an international airport, enclosing a large population – in the case of Heathrow, perhaps in excess of one million people (Sealy, 1976, p. 21) – and within such an area, as Harvey *et al.* document, the movement of modern aircraft constitutes a 'hazardous environment' (1979, pp. 263–4).

Secondly, there is the longer-term issue of the economic, or multiplier effect, of such a large activity. Studies undertaken of the impacts on the regional economies following the expansion of Shannon and Heathrow have indicated major growth-pole effects (Hoare, 1974; Tarrant, 1967). Sealy outlines the results of studies undertaken in New York, which suggest that upwards of half a million inhabitants might be expected to move to a region in which an airport is located, as a result of employment opportunities and related service development (1976). Foot relates similar projections in the British case, with a four-lane airport directly attracting some 65,000 jobs, with another 18,000 jobs being available in related sectors. This in turn would generate some 42,500 service jobs; all in all a net increase of over 300,000 residents. Using spatial interaction models within the context of a Garin-Lowry model, it is possible to show that due to NNI constraints and existing development, a TLA at Thurleigh would have resulted in widely distributed growth extending over several hundred square miles (Foot, 1979).

It is worthwhile emphasizing the enormous 'spillover' effects that emanate from an airport, because they underline the fact that the location chosen must be as far as possible the correct one. This, in turn, raises the question of goals. What is the correct location? Broadbent argues that the 'rational' approach, developed from neo-classical micro-economics is to argue that 'a planning project is worth supporting if some people gain and no one loses; the less restrictive Hicks-Kaldor principle suggests a project is worth supporting if the gainers from it can compensate the losers' (1977, p. 235). It is this approach that underwrites the use of CBA in the London Airport case, in an attempt to find the location where the losers can be compensated at least cost to those who stand to gain.

As Broadbent argues, CBA has been widely criticized on two grounds. The first is that relationships cannot be reduced to simplistic economic terms in which conflicts are reconciled and consensus can automatically result. This is a difficult argument to discuss realistically, as it is so sweeping in its implications. However, we can, I think, point to several instances, such as the costs that are applied to phenomena like historic buildings or wild-life, that show how shallow the idea of economic compensation may

be. Clearly, the use of insurance value to 'price' a unique archi-
tectural monument that is likely to be destroyed in the wake of
construction is to put an artificial price on scarcity – some might say
on the priceless (Bromhead, 1973).

The second criticism is technical in nature, and has been extensively discussed in the context of the TLA (Self, 1970; 1975). Hall
also makes telling criticisms of the ways in which the user-costs, i.e.
the transport costs to and from the TLA, were calculated within the
Roskill CBA (1971). It will be remembered that Maplin/Foulness
was militated against by high user-costs due to its relative inaccessibility: Hall argues that the calibration of gravity models is a
notoriously difficult procedure, and that the way in which users of
airports value the time spent reaching the terminal is only a
'guesstimate'. This leads on to a subsidiary issue, namely the
impossibility of incorporating all available evidence into the CBA.
In this instance reduced commuter bills from Southend to London
should have been incorporated as one spin-off from providing new
airport-related jobs at Maplin; but they were not. Equally, the noise
impacts of keeping some airports open, and the benefits from
closing others (Southend for example) were only touched upon.

A further issue concerning CBA can also be developed from this
theme, namely the problems of applying CBA to a large, regional,
multivariate problem. Broadbent argues that other techniques
(such as the Planning Balance Sheet) need to be invoked, because
they can incorporate long-term impacts on employment, income
distribution and so on. As we have seen, the impacts of an international airport are of this magnitude. He concludes that:

The CBA technique appears to fail just at the point where it might hopefully
be expected to be most relevant. It does so not only because its overall
conceptual underpinning is questionable, but because conventional micro-
economics is most at home in handling partial situations . . . rather than at
complex, strategic planning situations. (1977, p. 238)

Broadbent asserts that 'a more conceptually rigorous method' is
required to cope with such issues. In essence this is correct,
although it should be noted in passing that CBA can also be used
effectively to answer questions, not simply about the impacts of
externalities, but also about the individuals who may or may not
benefit from the location. To examine the spatial distribution of *net*
costs ignores the possibly differential impact of gains and losses.
Thus, for example, it is clear that those who own land around a
proposed airport, and those who make use of air travel, stand to
receive a different amount of indirect benefits from those who
simply have use-value of property and do not (or cannot) make use

of air transport. As Smith observes 'it is these inter-group (often inter-area) transfers that are so important in redistribution' (1977, p. 187).

In his discussions on this theme Smith moves beyond CBA, and considers the possibility of examining major planning impacts via the application of a variation of input–output analysis that can incorporate various welfare functions; in other words, the kinds of distributional effects already alluded to. Using such an accounting framework, it is possible for one to follow through the chain of effects resulting from a major planning development, not only in economic terms, but with respect also to the resultant 'level of living' for different social (or spatial) groups (Smith, 1977, pp. 191–7).

To summarize this argument, I think we can reasonably suggest that in this case at least, the likely outcomes of developing the TLA were not fully explored by the research teams advising the Roskill Commission. This is not to suggest that, for example, the regional growth issue was unknown to those in a position to make advice available; however, this issue was not systematically explored. More importantly, the *social* impacts, the question of who gains and who loses, did not emerge. In treating the location of such a major externality as a 'merit good' – a universally desirable piece of hardware that can be established with a few compensatory payments to those inconvenienced – the Roskill experience indicates that the planning system is over-keen to simplify the situations with which it deals.

The arguments I have outlined here concentrate upon the depth of understanding reached by those involved in the TLA history. It might be argued that the remit of the Roskill Commission (and indeed other studies, such as that which investigated Stansted in 1965) was not to go beyond the aviation issues, i.e. the decisions revolved about the *location* of, not the merits of, a TLA. However, one can also raise some criticisms of the general technical performance of those mounting arguments about future air-traffic pressures. Even in the basic area of predicting air-transport flows (on which, of course, all conjectures are predicated) the estimates have been shown repeatedly to be unsatisfactory. As Hall shows, the predictions made by the Department of Trade for passenger movements have steadily increased in line with increasing incomes, mobility, changing fare structures and so on. However, their predictions for the 'saturation date', after which a TLA is vital, have continually been passed without incident. This relates to some degree to their reluctance 'to deal with unspecified trends' (Shaw,

1979). In other words, a 'rigorous econometric approach' has been employed, which cannot, or will not, take into account central issues such as changes in aircraft size (fewer flights are needed, proportionately) or the increasing acceptance of off-peak travel.

More importantly, the British Airports Authority (BAA) itself employs more realistic estimation procedures, that make allowance for such 'intangible trends' (Shaw, 1979, p. 188). Equally interesting is the BAA's long-term preference for developing Stansted, i.e. not building a new site at all, but rather incrementally planning for growth at an existing site. This would not necessarily involve any less impact on the surrounding area in the long term, but it would certainly be cheaper, and perhaps even quicker. Most interesting is, however, the fact that different groups of professionals can produce such very different long-term strategies: one major inquiry has proposed a new site, whilst Stansted has twice now been tipped as the long-term favourite. (Those who find such inconsistency surprising might like to wrestle with the suggestion made by Dutch authorities that even Schiphol could reasonably become London's Third Airport).

The community

If we return to Figure 5.3, we can see that the planning system does not stand alone: it is assumed to be involved with the community on the one hand, and the political system on the other. Indeed, Batty's diagram simplifies the interrelationships to some degree, as the community can, in this sort of situation, not only participate in the planning system, but sometimes by-pass it altogether in an attempt to access the political system direct. This section briefly examines the activities in this vein of the various communities which felt themselves threatened by the TLA, before going on to consider their relationship with the political process, as it is manifest in the form of the planning inquiry.

First, it is perhaps necessary to outline a little of what constitutes a 'community'; in fact, the phrase 'community of interest' is perhaps more useful. Any examination of the lists of individuals and groups who give evidence to planning inquiries reveals a bewildering collection of interests. Sometimes, as with the mysterious figure who submitted to Mr Justice Parker at Windscale a copy of the *I Ching*, the motives remain hidden. In other instances the interests eventually become clear, as with for example the submission to the Roskill Commission from the Bishop's Stortford Season Ticket Holders' Association, which was worried about the possible effects of building the TLA at Stansted upon the rail services that its members used for commuting purposes. Table 5.2 outlines some of the individuals and groups who submitted evidence to Roskill.

Within the list it will be noted that several 'spatial communities of interest' exist. Again, this cumbersome phrase is in order because it is usually unrealistic to speak of 'spatial communities' without quali- fication. Communities – in the sense of spatial groupings with common goals – are rarely observed without the existence of some threat which causes other conflicts to be suppressed. As we shall see below, there seem to be some minimum requirements even then for a mobilization of interest to come about; property rights and articu- lation are obvious considerations, but the French experience seems to suggest that some wider political awareness may also be an important prerequisite (Castells, 1978).

Table 5.2 Groups submitting evidence to Roskill (partial list)

Bishop's Stortford Season Ticket Holders' Association
Board of Trade
British Airports Authority
Buckinghamshire County Council
Country Landowners' Association
Federation of Essex Women's Institutes
London Transport Board
National Farmers Union
National Union of Vehicle Builders
Port of London Authority
Scottish Economic Planning Council

Source: Commission on the Third London Airport, *Papers and Proceedings*, Volume II, 1969.

In the case of the TLA the depth of locally based opposition is well documented. Perman, for example, traces the campaign organized by the inhabitants in the Cublington (Wing) area, and notes that whilst Roskill cost in excess of £1 million, nearly £¾ million was spent trying to overturn the decision. Emotional battle- cries were coined, whilst public-relations firms worked more quietly on Members of Parliament. Punch magazine wrote of 'Cublington Fair', at which 'middle-aged Freeholders . . . parade with quaint hand-lettered signs . . . in the hope of averting the curse of Airport Bill' (Perman, 1973, p. 15). As he notes, however, 'we can call them a community, as the Roskill Commission did, because they were made into one by the common threat of the airport. Without that threat, they would scarcely have recognised themselves as having much in common' (Perman, 1973, p. 61). Indeed, an early episode in the life of the Wing Airport Resistance Association was a curious tussle with a local Labour MP over the unrepresentative membership of the leadership of this community group: not surprising, perhaps, in an area close to the famed stately home, Mentmore.

The whole issue of the way in which any group makes its voice heard is particularly interesting; as intriguing as the way in which different groups, in this case representing different spatial areas, compete with each other for attention and influence. In the Cublington case the local community presented one case, whilst Buckinghamshire County Council submitted evidence to Roskill separately. Although their ultimate intentions were similar, the reasons why they opposed an inland site were not, and reflected a different horizon, On a wider level, Bromhead notes that 37 MPs spoke in the 'Roskill' debate; 15 represented threatened constituencies, and 14 spoke against the TLA as it related to his or her area; all tended to promote the Maplin Sands site, which gives an added dimension to Peter Hall's phrase concerning planning as 'spatial chess'. In turn, of course, the Foulness/Maplin site had its defence committee, the 'Defenders of Essex'; curiously, a University of Essex study of this period noted that although only very small numbers of families were threatened in this area, they did in fact constitute a unique community (Abell et al., 1969; Bromhead, 1973). Even more noteworthy is the fact that on a wider spatial scale, the Essex County Council was in favour of Maplin on general economic grounds; this illustrates rather well the pattern of co-operation and conflict that exists at different spatial resolutions, and reminds us of what Pepper has termed the lifeboat principle:

Concern for environment and countryside is a localised and personal rather than a universally applied principle in third airport protest . . . The lifeboat ethic derives from Hardin's metaphor of a lifeboat adrift in a sea in which people are drowning. The lifeboat's supply of food is sufficient only to sustain those who are already in the boat. These fortunate people argue that because of the limited amount of food (resources) they cannot help any of the drowning into the boat. (1980, p. 177–8)

In other words, the protests against an inland site were based not upon an all-embracing environmental belief, but very simple (spatial) self-interest, an assumption that the beauties of the countryside were not capable of being shared with any outsiders.

The political system and the planning inquiry

I have already suggested that there are perhaps less links in Figure 5.3 than reality warrants: the accessing of the political system (in the widest sense) by communities is one consideration, and the two-way relationship between the planning system and the political system is another. Thus, although the former is able to operate in a functional and rational manner, it must often do so within certain constraints. This is pointed out by Bromhead, who argues that in terms of Lindblom's model of decision-making, the Roskill era was one of 'incremental change' and 'low understanding'; despite the enormous expense and numerical analysis involved, the debate was confined, by political constraints, to the very narrow issue of where to put the TLA, not whether to build a Channel tunnel, opt for regional airports and so on (Bromhead, 1973, p. 221).

When we examine the political process in detail, it becomes immediately clear that any decision-making process is subject to a multitude of influences, which are to varying degrees reconciled. This can be seen quite clearly in the context of large-scale planning inquiries, representing the arena in which the planning system, the political system and the community confront each other. (As we shall see in Chapter 6, some planning inquiries have achieved a certain national notoriety, in the field of transport planning especially. However, this should not mask their basic role, which is the reconciliation of *local* conflicts). As yet, there has been very little systematic analysis of the political context of phenomena like planning inquiries; studies like that of Grant (1977) tend to be anecdotal, and operate within the realm of party politics and chance occurrences. We can, however, place these histories within a far firmer framework, which relegates the unique to a subsidiary position, and highlights the implications of the events.

The first step here is to place an event like a planning inquiry into its overall role within the day-to-day functioning of society. In that sense it is part of the decision-making process whereby those in certain social positions exercise authority over others. The use of the term 'authority' is deliberate here, and is taken from the work of Dahrendorf, who is associated with a social theory which emphasizes not necessarily conflicts in the sphere of production, but conflicts (and the generation of conflict groups) relating to the exercise of authority (Dahrendorf, 1959). He writes that

authority is a universal element of social structure. In this sense it is more general than, for example, property, or even status. Authority relations exist wherever there are people whose actions are subject to legitimate and sanctioned prescriptions that originate outside them, but within social structure. (1959, p. 168)

The most important point concerning authority is that it is a 'zero-sum' concept, whereas wealth is not. As Dahrendorf points out, in advanced societies there exists a continuum, extending from poor to wealthy. Authority is however more usually a binary thing: 'a clear line can, at least in theory, be drawn between those who participate in its exercise in given associations and those who are subject to the authoritative commands of others' (1959, p. 170). It is to be noted that Dahrendorf emphasized 'given associations', and suggested furthermore that individuals' positions of authority might vary, depending on which context was examined: workplace, church, cricket club and so on.

Increasingly, however, we can identify groups for whom a lack of access to authority is cumulative, and increasingly these groups are becoming spatially manifest. A common feature of the inner city, for example, is the concentration of families who have little control of the most basic decisions within their lives, be they to do with housing, their jobs or their children's education. Nor should this be surprising in a society where access to public goods is controlled by numerous managers and gatekeepers, and where some of the basic economic forces, particularly those operating in cities, are serving to compound the concentration and segregation of different groups. Suburbanization and selective commercial redevelopment both serve to separate those who have some economic control of their daily existence, and those who have not.

A planning inquiry must be seen in this context. In some situations the inquiry will focus upon an area in which there exists a conflict group, created by problems relating to housing allocation or transfer, job redundancy or a lack of public goods. Such an area may be sub-urban, or even a large sub-regional division. In such cases an opportunity to take advantage of the inquiry to access a different aspect of the decision-making process may not be taken-up: 'oppression and deprivation may reach a point at which militant conflict motivation gives way to apathy and lethargy,' (Dahrendorf, 1959, p. 217). Conversely, the inquiry may be seen as the first stage in a threat to an area, perhaps directly economic, perhaps ultimately economic, as with some threat to property rights. In this instance conflicts may be intense, particularly where there are residents of high economic status, or those accustomed to manipulating authority in other contexts: 'dominant groups are correspondingly not so likely to be as involved in the defence of their authority unless their high socio-economic status is simultaneously involved' (1959, p. 218).

Using this kind of framework, spatial conflicts assume a sharper focus. Rather than simply suggesting that communities are in competition in some form of spatial chess, we can identify a chain of events:

PLAN → CONFLICT → CONFLICT GROUP → ACTION.

At any point this sequence may break down; the plan may not generate conflict: its 'authority' may be accepted; there may be no community of interest formed: action may not result. However, these possibilities are all open to investigation, and they take us beyond the anecdotal level of many other studies: (see also Kirby, 1981b).

Résumé

The Third London Airport history reveals the extent to which a spatial issue can impose major tensions upon society, particularly those groups directly threatened by its proximity; more importantly, it reveals how the planning process, which can be assumed to exist for some essentially beneficial long-term purpose, can instead create specific conflicts, and in consequence conflict groups which reject the authority involved in the implementation of that process. As Dahrendorf has argued (although not in this type of context), such cleavages are increasingly replacing conflicts in the work-place as 'strong issues'.

The TLA example has developed these, and other themes. We can continue this chapter with a more explicit study of another contemporary spatial issue, namely nuclear power, and the Windscale Inquiry in particular.

Nuclear technology: a spatial issue?

The question of nuclear energy has a contorted past. Whatever its original destiny, the development of nuclear technology has been closely tied up with the development of atomic weapons, and not surprisingly the whole issue of atomic energy has suffered from this connection. The result of this association of ideas is that at a time of changing energy sources, the 'nuclear debate' is based, not so much on the economics of production, but on the safety of the reactors and the problems of nuclear-waste disposal. Perhaps this is to be expected, although it is rare for these aspects of other energy sources to be considered; there is no lobby to restrict coal production in order to reduce the deaths of miners and the incidence of bronchitis, for example.

The safety issue is, of course, a spatial one: nuclear energy is only a potential threat if it is produced in some proximity to residential areas (and it is noteworthy that the Atomic Energy Authority has implicitly met this problem head-on by building reactors in relatively isolated locations).[1] An opinion poll conducted in the UK in 1977 indicated that approximately one-third of respondents would either oppose the construction of a nuclear plant within 10 miles of their home, or would leave the area (White, 1977). Since that date, noisy (although 'minority') demonstrations in Germany (*vis-à-vis* the Brokdorf nuclear scheme) and in the USA (following the Harrisburg, Pennsylvania accident) have indicated that this proportion is likely to grow rather than to diminish (Stott, 1980). Breach has argued that this communality of interest transcends the spatial issue:

'no longer is opposition of a not-on-*our*-doorstep-please variety the norm: it has given way to a generic international campaign – in keeping with other reformist missions such as those concerned with resource conservation, wildlife protection and pollution abatement. (1978, p. 73)

Whilst an international lobby may exist, fuelled by authors like Jungck (1958), we return, however, to the motivation of those involved, and fundamentally this is one of spatial access: a wish to live in areas, or to use areas for recreation, without fears of radiation or pollution. Again, the lifeboat ethic is under examination.

Windscale: the political system and conflict resolution

Although the Third London Airport history and the Windscale Inquiry have some similarities (not least of which is expense; the latter cost £2m, spent within a far shorter period of time), there also exist several interesting differences which allow us to develop some additional themes. The TLA inquiries were to do with the siting of a major (noisy) externality; the Windscale Inquiry had a far more subtle purpose. Although it was essentially a local planning inquiry, with a brief to consider the development of a nuclear fuel-reprocessing plant, the 100-day sitting became a surrogate for a public debate upon the entire nuclear energy issue; a debate, moreover, that could be manipulated within a legalistic framework, and one that could, by granting the planning permission for a processing plant, appear to give some stature to the whole nuclear future. As we shall see below, it has been described as a simple exercise in legitimation (Kemp, 1980).

The Windscale history began in 1975, with an approach from Japan to British Nuclear Fuels Ltd, requesting the latter to reprocess their spent nuclear material. The Cumbrian site has processed indigenous waste for two decades (and even some foreign waste); however, the size of the proposed contract necessitated an expansion of capacity on the site. This proposal immediately became public property (for the Labour Government, an irritating 'nuclear leak') and discussion quickly settled down around three main topics.

First, there was, and is, disquiet concerning the importation of foreign waste: a typical jingoistic cry was that Britain was about to become the 'world's first nuclear dustbin'. Windscale has its share of industrial accidents, but these are ruthlessly reported; one such leak occurred at the time of government consideration of the Japanese deal. On the other side of the coin, quite literally, was the financial issue: the reprocessing contract involved some £900m in orders.

The reprocessing question, of course, dovetails with the wider consideration of nuclear energy itself. Bugler reports disquiet over the fact that Japan was using nuclear fuel, but was not a signatory of the Non-Proliferation Pact of 1978. The increased movement of atomic materials provides a dual threat, some argue: first, from terrorist hi-jackings and subsequent ransom attempts, and secondly, from the long-term loss of civil liberties that inevitably follows from attempts to deal with the hi-jack threat. On the one hand, then, are these potential side-effects; on the other are the problems raised by the 'energy gap': the predicted shortfall between energy needs and extant, economic sources. It was at this time that rising oil prices first began to focus general attention on this phenomenon.

The third issue was related to the spatial impact of expanding the Windscale plant itself. British Nuclear Fuels (BNFL) predicted a need for 1400 new jobs, and local trades councils responded warmly to this. Conversely, those committed to the use of the Lake District and other outstanding areas of natural terrain were particularly opposed to the visual, and the more long-term and sinister dangers of radiation.

The way in which these three subconflicts were read by Peter Shore, Secretary for the Environment, and his ultimate decision to opt for an inquiry under the aegis of the 1971 Town and County Planning Act, provide a fascinating insight into the decision-making process. We can consider his reactions in the context of a behavioural model, namely Kasperson's model of municipal stress management, which is developed in a similar context in Barr's study

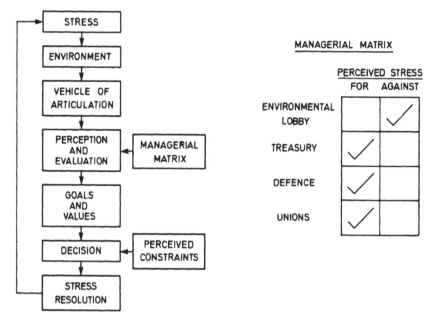

Figure 5.4   Kasperson's model of stress management (developed from Barr, 1978)

The model is presented in program form, and is to be read downwards. The 'stress' is in this instance the application by British Nuclear Fuels Ltd to reprocess more nuclear waste at Windscale, and the stress resolution is the planning inquiry which found in the appellant's favour. The managerial matrix simply illustrates the pressures upon the Energy Secretary to agree to BNFL's request, from the Treasury, the Ministry of Defence and unions supporting the jobs likely to result from the expansion.

of the construction of a (non-nuclear) power plant in Australia (1978).

In Figure 5.4 the model is outlined. Initially, stress emerges within the decision-making environment; that follows, in this instance, the announcement of the Japanese contract and its implications for Cumbria, resulting in environmental protests. These are well and widely articulated, and thus a serious threat to a government with a small majority in the Commons. This and other issues are assessed by those in a ministerial position, via the 'managerial matrix': in this instance several pressures are outlined, notably the 'anti-lobby' and the various factions in favour of development. The Environment Secretary is responsible for balancing the environmental interest with the wishes of the Treasury (considering the balance of payments), Defence (considering the atomic-weapons position) and the unions (considering employment). Goals and values clearly play a part here 'and in the end . . . the Windscale Inquiry took place because of public and private pressures, but most of all because a minister made a personal decision' (Bugler, 1978, p. 184). The important point here is that a local land-use planning inquiry resulted from this process, which reflects the arguments *against* a full 'nuclear debate', namely, that another Roskill, with all that that implies, would result. From a ministerial point of view, this meant that under the aegis of a local inquiry a basic planning submission could be examined, and objections raised. In such a situation the onus is very much upon the objectors to mount a satisfactory case, whereas in the case of a Royal Commission the onus of successful debate rests far more upon *all* those groups which are trying to influence decision-making. This argument is consistent with the view promoted by Kemp, namely that 'the public inquiry at Windscale, rather than enabling public participation in the decision-making process, merely served to legitimize state intervention' (1980, p. 366). It was thus the minimum exercising of the power of the state commensurate with the achievement of the goal in question: namely expansion of the nuclear-reprocessing industry.

Having considered the choice of an inquiry, let us now examine its deliberations. Once more we can usefully think in terms of 'the planning system' and 'the community', alongside 'the political system' that has already been briefly outlined.

The Windscale Inquiry and the planning system

In his review of Mr Justice Parker's report of the Windscale Inquiry, David Hall (Director of the Town and Country Planning Association (TCPA)) wrote, tongue firmly in cheek,

this slim (92 pages), over-priced (£3.75) volume by a previously little-known author in the field does not fulfil the promise which had been hoped for when its publisher, Peter Shore, announced the launching of the work at the end of 1976. (1978, p. 92)

On one level Hall alludes to the fact that the judgement made by Parker was unfavourable to the TCPA and the many other organizations and individuals who gave evidence. However, it also reflects a fairly widely held belief that the Inquiry as a whole was less than fair. As I have suggested, the choice of this type of inquiry was designed in itself to limit conflict, even though ultimately Parker allowed the hearing to roam far more widely than he need have done. This notwithstanding, unfavourable comments were made in relation to at least four aspects of the affair.

First, and most frequently mentioned, is the timing of the whole process. According to Bugler, 'pressure on time distorted the inquiry' (1978, p. 184). The Outer Circle Policy Unit (OCPU) links this aspect in with another issue, that of financial resources: 'the inquiry could not even be appointed without the "trigger" of a planning application. This meant that the proponents of the scheme had a head-start over the objectors in the preparation of their case, an advantage compounded by the disparity of resources between them' (OCPU, 1979, p. 24). They continue, 'the inquiry itself was taken at great speed, sitting five days a week in a remote part of the country, continuously for over five months without a break. This meant that the objectors did not have enough time to digest the information supplied by BNFL, much of which only emerged in the course of the inquiry' (1979, p. 24).

Other criticisms are also mounted. The legalistic nature of planning inquiries can deter some; however, the advantages of a legal *modus operandi* (such as prior circulation of documents) does not occur. Also, the role of the Inspector may be open to question: 'a single Inspector, even with expert assessors, presiding over a statutory planning inquiry is not the best way of achieving . . . a [credible] result . . . since it enables critics to blame an adverse outcome on the personality and alleged predjudices of the Inspector' (1979, pp. 25–6).

One of the criticisms of the Roskill deliberations was, by contrast, that the research upon which many of the conclusions of the Commission were based, was inadequate. As we have seen, the Windscale Inquiry differed in so far as it was not incumbent upon BNFL to discuss in detail wider issues, such as energy needs, within their application. None the less, such issues were raised by those in opposition to the application, raising again similar problems of estimation, the gap between the fundamental, even emotive issues at stake, and the technical language involved.

The issue of forecasting energy needs, within which context nuclear reprocessing, of course, must be understood, emerged as consistent with most attempts at prediction, i.e. some estimates made were wildly different from others. Some observers go so far as to state that 'many of the crucial questions raised at the inquiry are incapable of being answered factually, and to that extent it is hardly possible for the inspector to make a rational decision' (Smith, 1978, p. 407). In this sense it is possible to argue that the objectors to BNFL could *not* have won their case by argument of any type. By concentrating upon the debatable scientific issues (like the energy gap), they entered not so much a minefield as a morass: a sticky, grey area of conflicting statistics and extrapolations which could readily be turned aside due to their lack of accuracy, though not their precision (Conroy, 1978; TCPA, 1978). However, had they simply concentrated upon what I have called the emotive issues of nuclear energy, they would equally have been in a weak position; indeed, Parker concluded that 'much of the opposition appeared to be based on sincerely held moral grounds, and that amongst those who advanced opposition on such grounds there was a tendency to suggest that supporters were acting in an immoral way. This attitude is plainly unsustainable' (1978, p. 71).

Interestingly, this dilemma is well articulated in studies by political scientists, who note a basic tension between many opposition groups (whom they characterize as 'ideal-regarding') and planners and decision-makers (whom they characterize as 'want-regarding': Kirby, 1981b, pp. 240–1). In simpler terms, this is thus a clash of outlooks, with ideals opposing wants, which are frequently measured in objective (CBA) terms. This tension is reconcilable, but most objectors feel the necessity of trying to express their views in 'want-regarding' form, due to the all-pervasive technical nature of the planning process; the net result is usually a poorly articulated case.

In view of this it could be argued that the objectors might have been better advised to base their arguments upon *planning* considerations, i.e. to treat the inquiry as narrowly as its legal status warranted. It is clear from the Inspector's report, however, that given all other considerations (such as the need to reprocess fuel), the Windscale site is clearly the right one to use for this purpose, if only because it already undertakes such activity: 'in planning terms, there is no substantial objection to the proposed development' (Parker, 1978, p. 78).

The communities of interest

A point that follows from the type of case that the objectors raised, is the identity of those groups and individuals that were involved in the inquiry. As Table 5.3 shows, the broad nature of the cases argued reflects to a large degree the nature of their involvement in Windscale as a spatial issue, rather than a personal threat.

Indeed, Windscale represents a curious contrast to the TLA story, in so far as there were no locally based protest groups, with the exception of those societies with a prior existence (such as Friends of the Earth (West Cumbria) and the Cumbria Naturalists' Trust). From this we must infer that the long-term existence of a

Table 5.3 Partial list of Windscale Inquiry participants

Name	Interest/speakers	Viewpoint
British Nuclear Fuels Ltd	Applicant	For
Cumbria County Council	Planning	For
Copeland Borough Council	authorities	
Friends of the Earth	Ecological	Against
Isle of Man Local Government Board	Tourism	Against
Scottish Campaign to Resist the Atomic Menace	Self-explanatory	Against
Society for Environmental Improvement	John Tyme and Arthur Scargill	Against
Town and Country Planning Association	Environmental	Against

Source: Parker (1978) Volume 2.

smaller plant at Windscale had removed any perceived threat, and consequently the supposed advantages, such as employment opportunities, increased in importance.

This in turn raises a further point, concerning distributional effects. Normally, locals' needs are sacrificed to the needs of strangers, although some form of payoff may result. There are, however, no precedents for strangers attempting to stop the construction of an externality that functions against their interests (recreation), but in favour of local needs (employment). Because employment is in many ways a more direct input to well-being than recreation, one is inclined to the argument that the opposition of these groups should be reconciled in favour of local employment needs. We are, however, immediately faced with the problem of comparing the costs of creating jobs (in this case about £65,000 each) with the costs of possible radiation damage and the loss of spectacular views and scenery.

This is, as we have seen, virtually irreconcilable. More certain is the fact that the groups who did represent themselves at the inquiry fall to some degree into the categories sketched in above: namely conflict[2] groups who have emerged in response to a specific threat (in this case the general issue of the environment) and which in the main have the articulation and are skilled enough in the accessing of authority to travel to a distant venue and present a highly technical opposition case. It is significant that of the 148 witnesses who

YOUR HOME
A NUCLEAR TARGET?

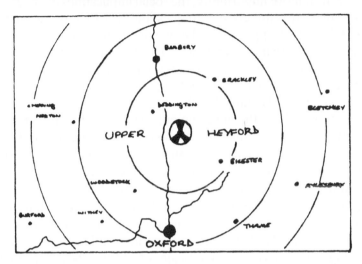

Figure 5.5   Extract from a leaflet issued by anti-nuclear group, Oxfordshire (Campaign ATOM, 1980)

The leaflet describes the horrors of nuclear war, but places them firmly within Oxfordshire. The leaflet goes on to state:

> If your home is in this area . . . The dropping of one 20-megaton nuclear bomb on Upper Heyford would destroy it. There is a plan to base 'Cruise' nuclear missiles at Upper Heyford USAF base; if this goes ahead, nuclear war becomes much more likely – a war in which Oxfordshire will be one of the first targets. 'Cruise' missiles form the latest stage in an arms race that is costing every man, woman, and child in this country £170 per year. It is the very fact that we have so many of these expensive nuclear weapons that makes us a target for attack. Surely it is time that we break the lunatic spiral, and make it clear that we don't want these missiles in Oxfordshire.

What the leaflet does not suggest is where the missiles might ultimately be located. Mention is made of the fact that 'Holland and Belgium have already said that they won't take these missiles', but no positive proposals are made that the missiles should be banned entirely.

presented evidence to Mr Justice Parker, 86 possessed degrees or some form of professional qualification.

This may be seen as an attempt to divert attention away from the strength of the environmentalist case, which clearly is a part of a growing international movement, in the sense that common aims and links exist. None the less, I would reiterate that opposition to nuclear power, and latterly nuclear weapons (by a revitalized CND movement), is again to be seen as an example of the lifeboat ethic, albeit of a rather more sincerely held nature. For all the discussion of long-term energy horizons, the environmental case is not radically concerned with the implications of changing energy patterns and, more importantly, the social implications of increased energy costs. It is not enough to question the extent of the energy gap (TCPA, 1978, pp. 20–3); the only certainties are that present energy sources are finite. Thus at *some* date, rising energy costs will have serious implications for societies with unequal income distributions, given the extremely limited potential of extant sources of alternative energy (Hoare, 1979).

Once more, then, the environmentalist case is diminished by its willingness to fight the nuclear issue in relation to specific localities, and to appeal to residents' self-interests – an approach extending even to anti-nuclear-weapons organizations, which deal in threats to communities rather than in pacifism (Figure 5.5).

To conclude this section, and to underline the singular nature of this particular planning inquiry, it is interesting to note that the spatial issues involved even led to some inter-union disagreements following the arrival at Windscale of Arthur Scargill, (then) President of the Yorkshire federation of the National Union of Mineworkers. Local union officials found his anti-nuclear, pro-coal sentiments 'lacking in sense and solidarity' (Breach, 1978, p. 113), and more likely to aid coal-rich Yorkshire than Cumberland, which has only one pit.

Conclusions

This chapter has ranged widely over a series of issues, and it is possible that several of the themes outlined here have become inter-linked. Some assertions – such as the criticism made of the nuclear opposition case – are controversial. Of particular importance, though, are the ways in which the planning process may be responsible for placing major externalities upon the landscape, and in relation to this, the ways in which individuals may have a say in this locational decision-making.

Some of the issues raised by Windscale and the Third London Airport are similar; the problems of prediction is an obvious example. In other ways the Roskill investigation and the whole, cyclical TLA saga are unique in the way in which the different communities have played space chess, with, it seems, an international airport as the booby-prize.

Although it may seem that this is its intention, this chapter is not designed solely to isolate the pitfalls of the major planning inquiry, although the OCPU's conclusions are that such inquiries do need to be radically overhauled, and replaced by a more flexible 'project inquiry' system (1979). Indeed, its observations are interesting in so far as it is recommended that there should be a separation of issues, wherever possible, between the national interest (an airport, a Channel tunnel) and the impacts upon specific sites and communities.

As we have seen, it is the impact upon specific localities that raises the greatest disquiet within the casual observer of the inquiry system. As these examples have shown, the right of speech does not guarantee a voice within the political system, or the planning system, for the community. This, indeed, is the main conclusion noted here: that unless certain conditions are right, no community, spatial or otherwise, will emerge; groups which are poorly placed with respect to authority will not re-create themselves as conflict groups prepared to access the inquiry machinery. Conversely, those used to accessing authority can achieve success in both the planning system and the wider political context. Moreover, when such groups do emerge, the strength of the resulting political cleavage, albeit temporary, is as intense, indeed often more so, than traditional cleavages relating to national 'political' issues.

Notes

1 More detailed research shows that an analysis of national population distributions could, however, indicate far more remote sites than those already chosen (Openshaw, 1980).
2 The extent to which such voluntary organizations as Friends of the Earth constitute a conflict group in the sense that it was used above remains to be seen; such a group is more likely to represent false conflict, in the context of the use of the term by social theorists such as Bailey (1975).

References

Abell, P. M., Bell, C. and Doreian, P. D. (1969) 'Papers and proceedings', *Commission on the Third London Airport*, 8(2), London, HMSO.

Bailey, J. (1975) *Social Theory for Planning*, London, Routledge & Kegan Paul.

Barr, L. R. (1978) 'Conflict resolution in environmental management; the Newport power station controversy', *Australian Geographical Studies*, 16, 43–52.

Batty, M. (1979) 'On planning processes', in Goodall, B. and Kirby, A. M. (eds) *Resources and Planning*, Oxford, Pergamon, 17–45.

Breach, I. (1978) *Windscale Fallout*, Harmondsworth, Penguin Books.

Broadbent, T. A. (1977) *Planning and Profit in the Urban Economy*, London, Methuen.

Bromhead, P. (1973) *The Great White Elephant of Maplin Sands*, London, Elek.

Bugler, J. (1978) 'Windscale: a case study in public scrutiny', *New Society*, 27 July, 183–6.

Castells, M. (1978) *City, Class and Power*, London, Macmillan.

Chadwick, G. (1971) *A Systems View of Planning*, Oxford, Pergamon.

Conroy, C. (1978) *What Choice Windscale?* London, Friends of the Earth.

Dahrendorf, R. (1959) *Class and Class Conflicts in Industrial Societies*, London, Routledge & Kegan Paul.

Foot, D. H. S. (1979) 'Mathematical modelling in land-use planning', in Goodall, B. and Kirby, A. M. (eds) *Resources and Planning*, Oxford, Pergamon, 51–76.

G. B. Commission On the Third London Airport (1971) *Report*, London, HMSO.

Grant, J. (1977) *The Politics of Urban Transport Planning*, London, Friends of the Earth, Earth Resources Research.

Hall, D. (1978) Review of the Windscale Inquiry, *The Planner*, 64(3) 92–3.

Hall, P. G. (1971) 'Stansted – an analysis', *New Society*, 28 January, 145–8.

Hall, P. G. (1977) 'Stansted – why not?', *New Society*, 13 January, 62–3.

Hall, P. G. (1979) *Great Planning Disasters*, London, Weidenfeld & Nicholson.

Harvey, M. E., Frazier, J. W. and Matulionis, M. (1979) 'Cognition of a hazardous environment', *Economic Geography*, 55(4), 263–86.

Hoare, A. (1974) 'International airports as growth poles: a case study of Heathrow Airport', *Transactions of the Institute of British Geographers*, 63, 76–96.

Hoare, A. (1979) 'Alternative energies: alternative geographies?' *Progress in Physical Geography*, 3(3), 508–37.

Jungck, R. (1958) *Brighter than a Thousand Suns*, London, Victor Gollancz.

Kemp, R. (1980) 'Planning, legitimation and the development of nuclear energy', *International Journal of Urban and Regional Research*, 4(3), 350–71.

Kirby, A. M. (1981a) 'A perspective on the British planning process', *Administration and Society* (in press).

Kirby, A. M. (1981b) 'Planning inquiries as a research issue', in Burnett, A. D. and Taylor, P. J. (eds) *Political Studies from Spatial Perspectives*, Chichester, John Wiley, 237–44.

McKie, D. (1973) *A Sadly Mismanaged Affair*, Croom Helm, London.

OCPU (1979) *The Big Public Inquiry*, London, Outer Circle Policy Unit.

Openshaw, S. (1980) 'Population distributions and the location of nuclear power stations', mimeo, Department of Town and Country Planning, University of Newcastle upon Tyne.

Parker, R. (1978) *The Windscale Inquiry*, HMSO, London. (3 volumes).

Pepper, D. (1980) 'Environmentalism, the 'lifeboat ethic' and anti-airport protest', *Area*, 12(3), 117–82.

Perman, D. (1973) *Cublington: A Blueprint for Resistance*, London, Bodley Head.

Sandbach, F. (1980) *Environment, Ideology and Policy*, Oxford, Blackwell.

Sealy, K. R. (1976) *Airport Strategy and Planning*, Oxford, Oxford University Press.

Self, P. (1970) 'Nonsense on stilts, cost-benefit analysis and the Roskill Commission', *Political Quarterly*, 41(3), 249–60.

Self, P. (1975) *Econocrats and the Policy Process*, London, Macmillan.

Shaw, R. (1979) 'Forecasting air-traffic', *Futures*, June, 184–94.

Smith, D. M. (1977) *Human Geography – a Welfare Approach*, London, Edward Arnold.

Smith, R. J. (1978) 'One hundred days at Windscale', *New Society*, November, 405–7.

Stephenson, L. (1979) 'Towards spatial understanding of environmentally based voluntary groups', *Geoforum*, 10, 194–201.

Stott, M. (1980) 'Nuclear confrontation', *Town and Country Planning*, 49(3), 91–2.

Tarrant, J. R. (1967) 'Industrial development in Ireland', *Geography*, 52, 404–7.

White, D., (1977) 'Nuclear power', *New Society*, March, 647–50.

6 Local conflicts

The rat danced up and down in the road, simply transported with
passion. 'You villains!' he shouted shaking both fists, 'You scoundrels,
you highwaymen, you – you – road-hogs!'
                                    (Kenneth Grahame: *The Wind in the Willows*)

Whereas nuclear plants and airports are threatening and emotive,
roads and homes are commonplace and even desirable. They
do however, function as negative externalities, albeit on a far
smaller scale, and their construction invariably involves the disruption of the urban fabric and the displacement of population. As
the case studies examined below indicate, the urban renewal
process can itself involve a great deal of planning blight, un-
warranted decay and hardship; this whole process is often sparked
off, in fact, by the need to improve road networks. Roads themselves, once relatively quiet thoroughfares, have become sources of
noise, pollution and the locations of injury and death. As the
quotation above indicates, the arrival of the motor car into the
Edwardian landscape was not a smooth process; indeed, as
Plowden shows in enormous detail, roads have for long periods
been centres of bitter legal battles based on the basic rights
of freedom of movement on the one hand, and peace and quiet (and
safety) on the other (Plowden, 1973). Nor is this process at an end;
Tyme has recently written that 'it is my belief, and the one shared by
an increasing number of people, that the motorway/trunk-road
programme with all its ramifications poses a consummate evil, and
constitutes the greatest threat to the interests of this nation in all its
history' (1978, p. 1). This is clearly hyperbole, but the impact of
such externalities is obviously not to be underrated, and it is with
this aspect that we begin.

Roads as externalities

The impact of roads upon households is described by Humphrey *et
al.* as follows:

if they reside close to a multi-lane limited-access highway with a daily traffic-flow of 56,000–75,000 cars and trucks, they are exposed to sound pressure levels of 85 decibels – a level that is 20–30 decibels greater than the sound created by traffic on lightly travelled streets. This residential environment reduces the audibility of speech, impairs communication, creates odours and dirt which diffuse throughout the area, and may cause behaviour changes among highway neighbours. (1978, p. 248)

Conversely, roads do have some social purpose, namely to facilitate an increase in accessibility to urban goods, workplaces or other parts of the city or region. The balance between these so-called 'spillover effects' (see Smith, 1977, pp. 114–16) has been summarized by Wheeler, as shown in Figure 6.1. The diagram reveals, as we might expect, that both costs and benefits decline with distance from the road, and that the levels of noise and air pollution decrease, as does the benefit of the road to those who cannot immediately access it. The important point is, however, that in proximity to the road, the environmental costs outweigh the economic and social benefits (it is also perhaps worth pointing out

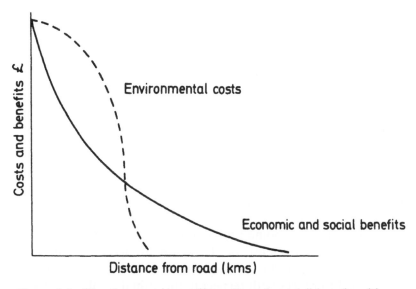

Figure 6.1   The distance–decay effect of benefits and disbenefits with distance from a road (source: Wheeler, 1976)

The vertical axis shows the way in which costs (such as falling house prices and double-glazing) and benefits (resulting from increased accessibility) vary with proximity to the road. It can be seen that near to the highway, the costs outweigh the benefits, whilst beyond a certain point (which will of course vary proportionately to the size of the road and the amount of traffic it carries) the benefits become more marked. In crude terms, the effect on house prices will be to concentrate disbenefits on lower-income households (but see also Figure 6.2).

that Wheeler does not counterbalance the economic advantages with the impact of the road on house prices). Overall then, a road constitutes a social cost for those nearby, and a clear gain for those some distance away. This of course may be compounded by car ownership, as Wheeler again observes; in his empirical study of the impact of an interstate freeway development in Atlanta, he shows convincingly the extent to which these benefits and disbenefits are perceived by the different communities affected by the development. Table 6.1 compares the responses of residents living in different neighbourhoods; Highland and Morningside are both in the path of the projected road, Oakdale is something in the region of a mile away, whilst Tanglewood is over three miles away.

Table 6.1 Mean responses to selected questions in four neighbourhoods: high values indicate agreement.

Questions relating to the proposed road development	Highland Morningside (close)		Oakdale Tanglewood (distant)	
1 Cause noise problems	2.6	1.4	2.0	3.0
2 Cause air pollution	2.7	1.8	2.1	3.4
3 Cause neighbourhood disruption	2.2	1.6	1.7	2.2
4 Cause neighbourhood blight	3.0	2.2	2.4	3.4
5 Improve access in Atlanta	2.5	3.5	2.6	1.8
6 Improve access to local services	3.0	4.2	3.5	2.4
7 Improve local social contacts	3.4	4.3	4.0	3.3
8 Reduce traffic congestion	2.8	3.9	3.4	2.3
9 Increase land values	2.7	3.9	3.2	2.3
10 Create benefits near route	3.4	3.8	3.2	2.3
11 Create better downtown	2.8	3.8	3.2	2.3

Source: after Wheeler (1976, p. 71) 1 = strongly agree   2 = agree   3 = no opinion
4 = disagree   5 = strongly disagree

Although there are some problems of interpretation in Table 6.1, (for example a response of '1' (agree strongly) and one of '5' (strongly disagree) when averaged indicate '3' (no opinion)), there are clear differences with respect to the attitude of the localities to the proposed freeway. The two neighbourhoods that are closest are clearly worried about pollution and noise, whilst the more distant ones are able to pick out the more advantageous effects, such as improved accessibility. The pattern is not solely a function of distance however: Morningside and Oakdale are both high-income areas, and show some basic similarity of response. This is, however, heightened by statistical analysis of the results; using principal components analysis, Wheeler was able to outline more detailed patterns of response (for discussion of this method, see Goddard and Kirby (1976)).

Table 6.2 Principal components analysis of responses to highway development: questions loading strongly on different components (see text for details)

Variables	Highland	Morningside		Oakdale			Tanglewood		
	I	I	II	I	II	III	I	II	III
Noise	✓	✓			✓		✓		
Pollution	✓	✓			✓			✓	
Disruption	✓		✓		✓			✓	
Blight	✓	✓			✓			✓	
Access–Atlanta	✓	✓		✓			✓		
Access–local	✓	✓		✓				✓	
Contact	✓	✓				✓	✓		
Congestion reduced	✓	✓		✓	✓		✓		
Land values	✓	✓		✓			✓		
Benefits	✓		✓			✓			✓
Downtown	✓	✓		✓					

Source: after Wheeler (1976, pp. 71–3)

Table 6.2 shows that Highland has only one pattern of response, which is one of opposition. Morningside residents also pick out the problems of pollution and so on, but in addition the short-term problems of disruption and the long-term problem of a lack of benefits accruing to that neighbourhood. Oakdale, in contrast, is able to separate the environmental problems from the overall benefits to be gained from a local freeway, as is Tanglewood. Both areas pick out the positive aspects (reduction in congestion etc.) as real gains to be made, and both neighbourhoods agree that the road will bring benefits to those living near the route.

This empirical example seems to bear out rather well Wheeler's argument, expressed in Figure 6.1. Unfortunately, we have no further insight into the situation observed in Figure 6.2, as all four neighbourhoods are relatively wealthy, and we can anticipate that car ownership is likely to be high. A fifth area sampled initially by Wheeler and located within a downtown black part of Atlanta, returned virtually no responses, and so attitudes to the freeway from a group of less likely users remain unrecorded; (this non-response, of course, is a predictable example of non-involvement, discussed in Chapter 5).

Roads and group protest

The Atlanta study suggests, not surprisingly, that the threat of a negative externality produces a mixture of group hostility and approbation, with proximity determining the reaction. How then do these groups react, and more specifically, how do they operate within the planning process?

A study in Washington, DC, of the construction of another interstate freeway noted a predictable process of assimilation by residents. Initially 'conversations' are typical: in other words, the process of attempting to create a community of interest begins. This tends to be followed by personal activities: 'steps' being taken to move house, which are normally abortive due to lack of demand and falling prices for property in the area; 'modifications' to the home, notably double-glazing follow; and finally, amongst those most disturbed by the road, 'complaints' to political representatives occur (Humphrey *et al.*, 1978).

This study suggests that the American experience may be different from the British norm on several fronts. The authors note that 'no indications were found of a challenge to the legitimacy of the highway itself' (Humphrey *et al.*, 1978, p. 262) which is, as we shall see, very different from the classic opposition case at many inquiries in the UK. An obvious source of differentiation is the fact that car ownership rates are far higher in many American neighbourhoods, and as Figure 6.2 predicts, 'residents who [are] heavy users of the highway tend to express less annoyance' (Humphrey *et al.*, 1978, p. 263). None the less, protest is not totally alien in the US experience: Mumphrey and Fromherz, for example, present detailed descriptions of the mechanisms of a transport-planning simulation, involving roles such as 'Percy Picket' and other familiar caricatures (1978, p. 285).

As suggested, the British experience is one of intense protest concerning not only motorways and trunk roads, but intra-urban improvements and even back streets. Victorian street plans, frequently of a rigid grid-iron pattern, are very susceptible to the creation of traffic-free precincts; indeed, bollards cemented across roads are a sure way of identifying a General Improvement Area (see Chapter 2). However, the creation of traffic-free roads also leads to the channelling of cars into 'rat-runs', at the expense of the residents of these particular streets, as noted by Cox (1979, p. 287).

Figure 6.3 illustrates a small part of such a traffic scheme undertaken in Jesmond, Newcastle upon Tyne, in 1970. Until that date Jesmond appeared to function as a relatively homogeneous

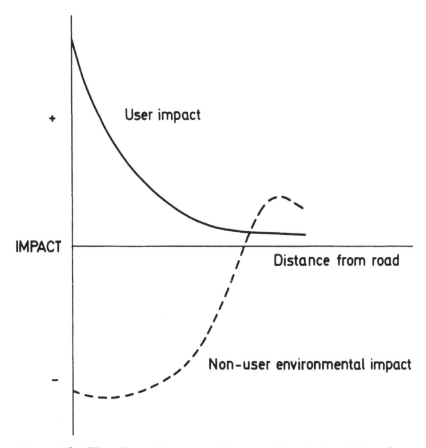

Figure 6.2   The effects of car ownership on residents in the vicinity of a highway (source: Wheeler, 1976)

The diagram considers positive effects (above the horizontal axis) and negative effects (below), upon both car users and non-car users. Users gain benefits (of accessibility), but these decrease with distance from the road. Non-users suffer only disbenefits in the vicinity of the road. Note that Wheeler suggests that away from the road, even non-users may however gain benefits: principally from a general increase in land values (resulting from increased access) and the removal of traffic from smaller roads onto the main thoroughfare.

community, represented by a single residents' association. In that year several roads were blocked off to create quiet residential precincts; this, however, coincided with other traffic closures on major routes, and consequently a good deal of central-area traffic became concentrated into a small number of east-west routes, such as Osborne Avenue. The latter increased its traffic flow by between 20 and 35 per cent, whilst adjacent Holly Avenue, lacking through access, experienced virtually a total absence of non-residential traffic.

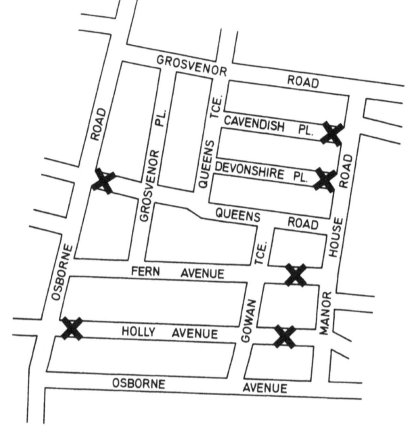

Figure 6.3   Road improvements, eastern Newcastle, 1970

The large crosses represent residential streets blocked off to through traffic. Holly Avenue gains particularly from an absence of through movement (quiet; safety) whilst Osborne Avenue suffers (greater volume of traffic directed onto this road).

The extent of resident reaction to this change was surprising. Both streets formed their own residents' associations: Osborne Avenue to fight for both roads to be opened, Holly Avenue to campaign for both roads to be closed. Residents of Osborne Avenue carried out traffic surveys, and launched a protest campaign based on the environmental problems facing a street containing a hospital, three old-people's homes and six doctors' surgeries. Ultimately a local referendum was organized by the city council, in which 98 per cent of Holly Avenue residents voted for maintaining the scheme, whilst 93 per cent of Osborne Avenue voted against; and throughout the area as a whole, 391 wanted the scheme to be retained, whilst 792 wanted it changed or scrapped. As a result, most of the streets (with the exception of Holly Avenue) were reopened. However, despite this success, three local councillors ultimately lost their seats, after long representation, at the next ward elections.[1]

The question of the political process has been examined at a rather different scale by Grant, who has documented the traffic plans developed in three British cities (Southampton, Portsmouth and Nottingham). His conclusion is that large-scale developments (such as Nottingham's famous 'collar scheme') can readily become political footballs if the area is at all politically marginal:

In Nottingham, the range of views and the contributions of the protesters were diverse. The Chamber of Commerce and the Civic Society emphasised the damage that the Eastern By-Pass and Sherriffs Way would do to business concerns, and to the physical structure . . . . St Anne's Tenants and Ratepayers Association demonstrated the problems that these roads would produce . . . . The Bus Users' Association, the Labour Party and the University academics provided positive and valuable contributions to an alternative policy. (1977, p. 138–9)

Clearly, in such a context, a spatial issue can be sucked into the 'normal' political situation. As the following sections illustrate, however, many schemes, particularly those relating to motorways or trunk roads, evolve as 'apolitical', but intense local conflicts.

Roads and conflict: a case history

Although it does not deal with an area of outstanding natural beauty or a venerable town, the background to the proposed link between the M1 and A1 in West Yorkshire reveals an interesting series of proposals, conflicts, claims and counter-claims.

The first moves date back to 1975, when transport affairs were still controlled by the Department of the Environment. In that year the DOE circulated to residents within West Yorkshire a detailed leaflet, which outlined the nature, costs, directions and numbers of properties to be destroyed by four different routes connecting the M1, which terminates in Leeds, to the A1, which constitutes the only major routeway from England into Scotland. The four routes, Red, Brown, Purple and Blue, are illustrated in Figure 6.4, along with some of the information provided to facilitate public choice.[2] In addition, recipients were asked to complete the questionnaire reproduced below.

Although 60,000 leaflets were distributed in both east and west Leeds, only 8000 respondents and 150 organizations returned their questionnaires; a two-to-one majority of individuals was in favour of an eastern route, whilst a majority of organizations and local authorities was keen upon a western route. William Rodgers, Minister of Transport, announced his choice as the eastern, Brown route (see Figure 6.4) in 1977, emphasizing the environmental damage to the Wharfedale area likely to arise from either of the western route proposals.

Within two days protest groups began to emerge. With a statement that the proposal constituted 'nothing less than the rape of the two communities', a parish councillor for the villages of Barwick-in-Elmet and Scholes (Leeds) urged a unified campaign of protest, despite the fact that, in the words of a local journalist, 'the people in Barwick and Scholes will tell you that there has never been much love lost between the two villages'.[3] In the local elections in September 1977 an Ecology candidate standing for the Wetherby ward based his campaign on the slogan 'another road?', whilst the East of Leeds Motor-route Action Group (ELMAG) also came into existence, its aim being 'to link all threatened villages along the route into a single, strong organisation to gather and evaluate the vast amount of material needed in order to present a strong and viable case at the public inquiry to be held in about two years time'. ELMAG's publicity thundered 'this road is on your doorstep – it will endanger the lives of your children – it will affect the value of your property – it will destroy our beautiful countryside – "industry follows the road", it will bring noise and pollution to our village'.

Towards the end of the year the West Yorkshire County Council's transportation committee decided in favour of an eastern route: not the DOT's choice (Brown) but the Red route; (the

Figure 6.4   Information provided to residents in Yorkshire, relating to proposals to link the A1 and M1 routes

(a)   A large document provided various items of information to households, in order that they could fill in the questionnaire and express their views on the most desirable route for the link.

(b)   A map showing the A1 and M1 routes, plus the four possible connecting corridors: each was identified by a colour code. Two principal routes were suggested: one to the west of Leeds (favoured by industrial concerns) and one to the east (favoured by environmentalists).

(c)   The environmental and construction costs of each corridor are shown in the data table; there are large variations in the costs of construction, the amounts of land required, the numbers of homes to be demolished, and the numbers of homes to be affected by noise and thus requiring sound-proofing.

Despite a public inquiry in 1979, no decision has yet been taken on the ultimate route for the link, and no date exists as yet for the commencement of construction.

Labour opposition favoured the Blue (west) route). The rationale for this choice was that fewer road improvements would be needed, with an overall saving of nearly £12m. This decision was the first of several announcements concerning the detailed routing of the eastern corridors. Early in 1978 Boston Spa, Walton and Thorp Arch parish councils reacted against suggestions made by nearby Wetherby Action Group, that the Brown route should pass through

(a)

QUESTIONNAIRE
YORKSHIRE TO THE NORTH EAST
KIRKHAMGATE - DISHFORTH ROUTE

Please answer the questions below by ticking the appropriate boxes.

1. Bearing in mind the needs of the Region and the country as a whole, which THREE of the following factors do you consider most important in choosing a route?

    i Effect on agriculture .. .. ☐

    ii Effect on landscape, amenity and recreation .. .. ☐

    iii Keeping costs to a minimum .. .. ☐

    iv Demolition of property .. .. ☐

    v Visual effect on property close to, but not physically affected by, the proposed road .. .. ☐

    vi Effect on shopping, schooling and community life generally .. .. .. ☐

    vii Noise and pollution .. .. ☐

    viii Effect on traffic e.g. reduction in congestion or journey time .. .. .. ☐

    ix Inconvenience and disruption during construction .. .. .. ☐

    x The removal of long distance lorry traffic from populated areas .. .. ☐

    xi Improving Industrial opportunities .. .. ☐

    xii Any other factor (please write here) .. .. ☐

	RED	BROWN	BLUE	PURPLE
2. Which would be your first choice?	☐	☐	☐	☐
3. Which would be your second choice?	☐	☐	☐	☐
4. Which would be your third choice?	☐	☐	☐	☐

5. Do you think your home or any property owned or occupied by you would be adversely affected by any of the alternatives?   YES ☐   NO ☐

	RED	BROWN	BLUE	PURPLE
6. If the answer to Question 5 is 'YES' please indicate by which routes.	☐	☐	☐	☐

7. Any other comments:

8. Please write your name and address

When you have completed this form please post it to reach us by
1 AUGUST, 1975. THANK YOU.

(b) Figure 6.4 (cont'd)

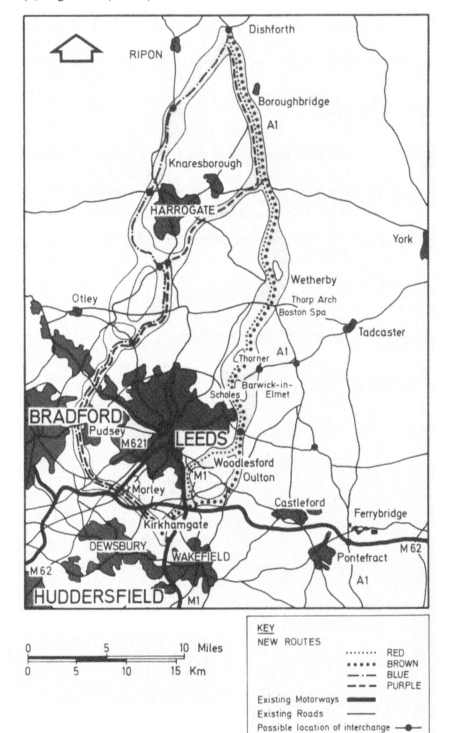

(c)Figure 6.4 (cont'd)

	Red	Brown	Blue	Purple
Probable standard of road	New dual two-lane high-capacity road plus improve-ment of A1, where neces-sary, to dual three-lane all purpose road (or equivalent capacity road) with full control of access and flyover junctions. An additional single all-purpose carriage-way would be constructed adjacent to A1 where necessary to cater for local traffic	New dual two-lane high-capacity road plus improve-ment of A1, where neces-sary, to dual three-lane all purpose road (or equivalent capacity road) with full control of access and flyover junctions. An additional single all-purpose carriage-way would be constructed adjacent to A1 where necessary to cater for local traffic	New dual two-lane high-capacity road	New dual two-lane high-capacity road plus improve-ment of A1, where neces-sary, to dual three-lane all purpose road (or equivalent capacity road) with full control of access and flyover junctions. An additional single all-purpose carriage-way would be constructed adjacent to A1 where necessary to cater for local traffic
Approx. route length from Kirkhamgate to Dishforth	58 km (36 miles)	58 km (36 miles)	67 km (42 miles)	69 km (43 miles)
Estimated construction cost of main scheme	£59 m	£62 m	£80 m	£76 m
Estimated construction cost of supplementary highway schemes	£25 m	£25 m	£13 m	£17 m
Total capital cost	£84 m	£87 m	£93 m	£93 m
Allowance for abnormal traffic delays during construction	£6 m	£6 m	£2 m	£3 m
Total capital and delay costs (1974 prices)	£90 m	£93 m	£95 m	£96 m

continued overleaf

	Red	Brown	Blue	Purple
Approx. area of land in hectares, required for main route (supplementary route)				
(a) agricultural land	200 (90)	200 (90)	260 (30)	260 (40)
(b) non-agricultural land	20 (50)	20 (50)	30 (5)	30 (5)
(c) Total	220 (140)	220 (140)	290 (35)	290 (45)
Approx. number of houses likely to be demolished				
(a) for main routes	30	30	40	40
(b) for supplementary routes	95	95	5	20
(c) Total	125	125	45	60
Approx. number of houses within 100 metres of highway boundary likely to need protection from noise by earth mound, etc.				
(a) for main routes	300	300	350	300
(b) for supplementary routes	1540	1540	310	440
(c) Total	1840	1840	660	740

their communities: not surprisingly they suggested that the route should pass through . . . Wetherby. Leeds City Council also voted for the Red route at the same time, and although ultimately both the County and the City reluctantly accepted the Brown proposal, North Yorkshire County Council maintained, until August 1979 a preference for the Blue route on the grounds that it would constitute a by-pass for Harrogate and Ripon (see Figure 6.4).

One of the more interesting aspects of these proposed improvements to communications in the Yorkshire region is the multiplicity of alternatives and the consequent plurality of support. In addition to the local authorities already outlined, industry was vocal in support of a westerly route, through the medium of the Blue Route Action Group (BRAG), which claimed to represent in excess of 19,000 companies, chambers of trade and commerce from Manchester, Sheffield and Tyneside, the CBI, the Road Haulage Association, North Eastern British Road Services and the Yorkshire Brewers. An interesting innovation is the intervention of the Royal Automobile Club; this is the first time that the RAC has pronounced on such a situation.

Despite this arsenal of supprt, it must be noted that BRAG only came into being in its support of the westerly route in February, 1978. However, the Pudsey Motorway Action Group (PUDMAG), that *opposes* the Blue route, existed in 1975. Indeed, the perceived threat precipitated a close involvement by PUDMAG in the distribution of the DOE questionnaire of that year, to the extent that the group distributed its own copies of the official form when these became scarce. When this became known three years later, it sparked off a major row between communities in the east, represented by ELMAG, and in the west, represented by PUDMAG. ELMAG claimed that PUDMAG had unfairly influenced the outcome of the 'referendum', and protested not only to local Members of Parliament, but even to the Minority Rights Group of the United Nations based in Geneva.

Roads and conflict: general principles

From Pudsey to Geneva is a long way; the depth of feeling on the one hand; and the sophistication of protest on the other, indicate that the community groups and interest groups involved in these struggles are sincere, motivated, well organized and above all, well financed. A leaflet distributed by ELMAG reveals that within a 15-month period, the small village of Thorner alone collected £3027, whilst ELMAG as a whole collected in excess of £7000. The group possesses a technical committee, composed of 'several engineers, surveyors, a geologist, a town planner, a mathematician and a physicist'. In 10 months, the committee carried out:

   (i) a critical analysis of DOT traffic flow data;
  (ii) collection of original traffic flow data;
 (iii) site assessment for various alternative routes;
 (iv) preparation of technical briefs for meetings held by ELMAG with district and county councillors, Members of Parliament and Ministry of Transport representatives and for associated press statements.

The group profile outlined above equates well with Twinn's findings in a study of the M3 Joint Action Group (M3JAG) in Winchester, which are summarized in Table 6.3; there is a clear distinction between those involved in the organization of protest, and those whose interests are represented.

Table 6.3 Comparison of M3JAG and Winchester population

Attribute	Winchester* %	M3JAG %
Sex (M)	38	79
Social Class I, II	39	93
Education >17	29	93
Age ⎧ 17–35	26	25
⎨ 36–55	38	53
⎩ 56+	36	21
Home ownership	44	75
Driver	49	93
Residence >5 yrs	63	68

Source: after Twinn (1978).
* Based on a sample of 205 residents.

Although the data on Winchester residents may be subject to some error, the M3JAG profile is clear: male, middle-aged homeowners and car owners, typically from the managerial and professional classes who have lived in the area for some years.

With this type of individual involved, it is not coincidental that Winchester has been the site of one of the bitterest, and it seems, most successful, community campaigns (the M3JAG has spent in excess of £45,000 between 1973 and 1979). Of particular importance is the fact that it was at Winchester that opponents of the construction of a road were first able to present a case based on the need for that road, as opposed to the more straightforward issue of the route of the highway. In his submission to the Inquiry Inspector, the ubiquitous John Tyme stated:

that at the 1971 Line Order Inquiry objectors had been denied the right to object to the motorway on the grounds of need; accordingly the inquiry should be adjourned and the matter of need should be the subject of a fresh inquiry. (1978, p. 37)

Of course, to undertake such dedicated protest requires not only cash and commitment, but expert analysis. The Winchester group hired a firm of transport consultants, who were able to show that the Road Construction Unit's figures on costs were awry (by £2½ million with respect to draining some sections of the proposed route) and that traffic forecasts were wildly unlikely (60 per cent increases by 1995, as opposed to the consultants' estimates of 32 per cent).[4] If a community can base a case around these types of issues, it seems that from their point of view a successful outcome is far more likely than in the past (as we shall see in the next section).

This latter point is well illustrated by the A1-M1 example, where the existence of 'salami politics', with each community fighting over a slice of the road at a time, means that any solution is a zero-sum gain; at the simplest level, if PUDMAG gains, then ELMAG loses. Both groups and both communities can only gain if *no* road is constructed, and this can only be achieved following a rejection on the grounds of need.

In some situations this becomes hopelessly confused, mainly for the reasons outlined in Figure 6.2, i.e. the existence of car users and non-users. In the Petworth By-pass Inquiry, which lasted in various forms between 1973 and 1977, two main possibilities were proposed as routes for the roads that at that time passed through the town (A283, A272, A285). One placed the by-pass through the National Trust property of Petworth Park, the other through a locally used beauty spot of Shimmings Valley. In terms of community needs this was a choice between an amenity rarely used and one (the Valley) frequently used; expressed another way, the choice was between improving facilities for through-travellers and motorists visiting Petworth Hall (with its 'historic' gardens) and taking account of local wishes, in an area of relatively low car ownership (approximately 42 per cent). In a parish poll in 1973, 99 per cent of those voting agreed that Petworth should have a by-pass; 89 per cent also felt that the road should go through the Park, and modified plans along these lines have been proposed by West Sussex County Council (Twinn, 1978, Chapter 5).

In addition, of course, there are also situations in which very localized groups are split over the route of a road. This was certainly the case with respect to the A1 Archway improvements in North London, which brought forth both an opposition case and a number of proponents. The fact that the latter were in the main local residents, and the former included individuals with what we might, in the widest sense call an ecological-cum-political, as opposed to a spatial interest, suggests that in some instances the business of protest may have a life, ideology and momentum of its own. It may also suggest that residents have as much to fear from those opposed to roads, even if they themselves are in favour of them in a particular case, as they do from the Department of Transport itself. Of these residents, John Tyme patronizingly writes,

the almost indescribable hell that motorway-generated traffic – chiefly juggernauts – has created for some local residents has ensured a vociferous, if not numerous lobby in favour of the road. I understand their anguish, but in that they are helpless pawns in the brutalised campaign to shift freight from rail to road, and in that the road widening would simply push the problem onto others . . . . I have always sought to convince them that their problem could be and should be solved by other means. (1978, pp. 64–5)

Clearly, the issues generated by spatial questions are subservient, as far as outsiders are concerned, to wider, aspatial principles (until the two, of course, coincide).

The political process: public inquiries into road proposals

The previous sections have given some flavour to the public inquiries set up to consider various road schemes, although we have not examined in detail the more vociferous aspects of the stage-managed protest that came to be associated with these events in the 1970s. The reasons why this depth of feeling existed is clear: in keeping with our former examples, threatened groups of citizens have been turned, through conflict, into forceful 'Motorway Action Groups'. Further, the source of the violence is born out of a feeling that capable preparation and forceful argument is not resulting in what we have termed an accessing of authority. Adams has described an inquiry as follows:

[it] is a one-sided contest. The promoter of the contest proposes the scheme and is also sponsor of one of the teams. He draws up the schedule and hires the stadium, which happens to have a dressing-room for his team only. He lays out the pitch on steeply sloping ground, with only one goal at the lower end, which happens to be the end his team is shooting at. He draws up the rules and brings his own referee. He equips the referee with 'Notes for Guidance' which insist that the ref. should on no account allow the opposition team to discuss any of these conditions, especially the absence of a goal at the upper end of the pitch. All appeals to the referee concerning the relatively trivial matters to be discussed must be couched in language which assumes that the promoter's team is always right. (1977, pp. 548–9)

In specific terms there are four things that have tended to provoke dissent at inquiries. As the 'Notes for Guidance' stress, it is not a public meeting: 'the Inspector must keep order and will not allow interruptions';[5] in other words, only those with the foresight, time and ability to have arranged to speak beforehand may do so. Secondly, the costs of attendance, and/or the costs of counsel, have not been refundable; many inquiries last for several weeks, at enormous aggregate expense. Thirdly, the issues of the need for a road have been sacrosanct: 'the Inspector may disallow questions to Departmental representatives which in his opinion are directed to the merits of Government policies'. Fourthly, in keeping with most people's feelings of helplessness when confronted by legality and bureaucracy, is the belief that the Inspector makes the decisions and that his mind is already made up (although in fact he does not: 'Ministers decide' (Department of Transport, 1978)):

the official title of the scheme being inquired into at the Archway is 'The Archway Road Improvement'. An objector put it to the inspector that a project which would knock down a large number of trees, shops and homes and drive a motorway through the middle of a community, was not necessarily an improvement. He asked to have the title changed on the grounds that it anticipated a verdict on what was supposed to be an open question. The inspector was unmoved. The word 'improvement' had become 'colloquialism', he said. It 'meant improvement for better or worse'. (Adams, 1977, p. 550)

The dissent shown at Winchester and elsewhere led to the (Leitch) Advisory Committee on Trunk Road Assessment, a new White Paper Review of Inquiry Procedures (April, 1978) and new 'Notes for Guidance'. The White Paper stated the following: that transport policy at the national level would be outlined, and 'opportunities will be provided for objectors to question the Department on these matters'; that greater liaison between the DOT and the public would be attempted, including the publication of alternative routes; greater informality wherever possible; some flexibility in allowing objectors to speak; no change in the payment of costs. It would appear that the Government was keen to ensure that 'an equitable and efficient public inquiry procedure' came about. Whether of course these changes bring this about remains still to be seen, due to the virtual collapse of the trunk-road building programme.

The planning process: forecasts and predictions

John Tyme's inspiration of attacking the very need for a road, as opposed to its location, echoes the arguments that have been used against nuclear power rather than the debate that surrounded the siting of the TLA. In both the latter cases, as we have seen, critical argument has centred upon the forecasting of future situations, and this is equally true of the prediction of traffic flows.

Given the possible sources of variation, it should not surprise us that transport planning can be criticized as wildly inaccurate. Lassière has outlined the parameters that need to be considered with respect to the siting of a new trunk road:

  (i) Specification of the characteristics of the study area; this includes projections of the number of residents, employment, land use and car ownership;
 (ii) forecasting of travel demand; this involves the development of a set of models, their calibration and use with respect to a future date;
(iii) specification of the base; this requires predictions of the operation of the highway and public-transport system that would exist in the plan year, if no further investment were made in transport facilities in the study area;
(iv) development of options for testing; options for testing are developed and selected from a larger number of possible combinations of road schemes, traffic management proposals, public-transport improvements and parking and public-transport policies.
 (v) evaluation of alternatives; economic, operational, environmental;
(vi) evaluation of the preferred plan; this may well consist of elements taken from more than one of the options tested. (abridged from Lassière, 1976, p. 3)

One or all of these factors may be liable to variation; moreover, roads are not readily built. As the Leitch reports notes, it may take from 7 to 15 years from inception to completion of a route; consequently predictions of demand must necessarily deal with the medium term rather than the immediate future (Leitch, 1977). Over such a time period, inspired guesswork is as common as extrapolation, and it must be said that past assumptions have not revealed a great deal of inspiration. For example, the macro-predictions created by the Transport and Road Research Laboratory in 1975 assumed an increase in national wealth of between 2 and 4 per cent per annum, and fuel costs increasing by between 0.0 per cent and 2 per cent per annum; since that date, economic performance in the UK has only held steady, and petrol costs have moved ahead in line with inflation.

At the local level it is often overlooked that the purpose of a construction, especially of trunk road status, is the saving of money:

making journeys, for whatever purpose, involves the road users in certain costs. These include the costs of time spent travelling, costs – both tangible and intangible – arising from road accidents, and direct costs of fuel, vehicle maintenance and depreciation. The building of a road will enable some of these costs to be reduced. It may reduce the travel time used to make a journey, accidents in the road network through diverting traffic to safer roads, operating costs, or all three. (Leitch, 1977, p. 34)

In order to justify construction, or thereafter a particular route, it is usual to apply CBA to a series of data, including the existing traffic flow, predicted traffic flow, vehicle-operating costs, accident rates on the present and future networks, and the travel time saved by the introduction of the new route. There exist clearly documented methods of achieving these figures: the costs to an employer of various types of travellers have all been estimated for example, around an average of 333 pence per hour (1976 prices). On the other side of the figurative coin, however, lie the indirect costs of construction, such as the demolition of homes, the increase of noise and atmospheric pollution experienced by remaining dwellings, the loss of agricultural land, and the numbers of accidents experienced by non-motorists. These shadow costs are *not* standardized, and despite their importance may thus not figure dramatically within the CBA. It will not have been forgotten, however, that these are exactly the issues that arouse public protest.

These criticisms are hardly original, as a reading of the (Leitch) *Report of the Advisory Committee on Trunk Road Assessment* indicates. Importantly, however, while emphasizing the desirability of a quantitiative approach, the *Report* observes that

whilst current methods of scheme appraisal, based on COBA [CBA] are sound as far as they go, we believe the assessment to be unbalanced . . . it is unsatisfactory that the assessment should be so dominated by those factors which are susceptible to valuation in money terms, and we believe it to be inadequate to rely simply on a checklist to comprehend environmental factors. (Leitch, 1977, p. 206)

The Committee is correct to stress that far greater attention should be paid to the indirect costs of demolition, and the various increases (and decreases) in noise pollution that occur in various neighbourhoods. What they do not, however, go on to link up are the issues that we have examined above, namely the trade-off that can be made by car owners between benefits and disbenefits, and the fact that this trade-off cannot be undertaken at all by residents in some low-income neighbourhoods. Clearly, it would be extremely difficult to work out the spatial variation in net gains and losses throughout the possible sites to be taken by a trunk road, not least because it would totally invert the process of route selection. Rather than assessing the balance of costs between doing nothing and building for any particular route, it would be incumbent upon highway engineers to identify low-cost solutions that picked out routes where gains exceeded losses for *residents*, rather than *users*. In a case like that stylistically represented in Figure 6.5, the choice would be between two by-pass routes, both costing similar capital

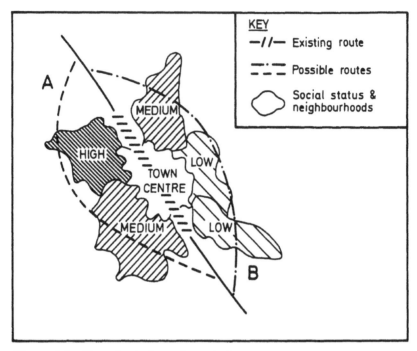

Figure 6.5   Hypothetical routings of a by-pass

The simple diagram is designed to show how different criteria might influence the direction of the by-pass. To the west, the route would cut through high-status housing, costing a good deal by way of compulsory purchase. To the east, the route would cut through cheaper properties. However, the levels of car ownership will be higher to the west, and the disbenefits and the benefits restricted to the same group of residents.

sums, but each offering a different set of gains and losses. In the above example scheme A might be desirable, as the increased capital costs of demolishing expensive homes are offset by the long-term increases in accessibility offered to car owners, who are more numerous to the west of the town. Although this may seem economic nonsense, the alternative is scheme B, which might involve cheaper compulsory purchase, but also offers no net gains to a neighbourhood with low car ownership. In terms of social justice, this would be a moral nonsense.

Although the situation presented in Figure 6.5 is hypothetical, it is not, in fact, unusual. Redevelopment, i.e. clearance, within inner urban areas for example, has frequently followed the planning of urban motorways, such as those in Birmingham, Leeds, London and Newcastle. The criticism that the DOE (and latterly the DOT) 'speaks with the mouth of the road engineer' is also seen in relation to local authorities in this type of situation, wherein comprehensive plans to move communities, redevelop large areas and build large numbers of new homes, are all predicated upon an external consideration such as a road. Malpass has recently argued that even prestige projects like the Byker redevelopment – which has been praised for its scope, vision, and often wrongly for the degree of public involvement – has stemmed in the main from a road development:

the corporation . . . was committed to redevelopment in Byker and to certain elements (such as the Shields Road Motorway) long before retaining the community was accepted as policy. These prior commitments, supported by well-established lobbies, ensured that retaining the community remained a low priority. (1979, p. 1012)

As Dunleavy observes,

'the large construction firms have been the most active section of the British Roads Federation in influencing local authorities towards urban motorway 'solutions' to traffic congestion; it is not coincidental that cities which achieved very close relations with contractors on public housing (such as Birmingham, Leeds or Glasgow) should have undertaken substantial urban motorway or ring-road programmes with the same firms. (1980, p. 125)

Urban renewal and the planning process

The example of Byker brings us, not entirely accidentally, to the issue of urban renewal, with which we may conclude this section. Due to the sheer scale and ubiquity of the renewal process within most British (and many European) cities, there are many well-documented studies of the impacts of this type of public-planning decision upon small neighbourhoods, and it is from this literature that we are most likely to gain some insight into the overall pattern of success and failure of spatially based protest. Moreover, this context links in well with the kinds of concern examined in Chapter 1, with an emphasis placed upon consumption. As Cockburn notes:

we are used to talking about struggle at the point of production. We know what it means – it is easy to envisage the factory floor, the building site or the office. One reason it is so easy to see is that it nearly always implies a place. The point of collective reproduction is not so easily pinned down. In some of its aspects it is of course a place too: housing takes place in a distinct location, and a council estate is a territorial unit. But private tenants renting from the same landlord may live miles, even cities apart . . . social casework and social security are delivered to people as individuals, sometimes at home, sometimes across a counter. (1977, p. 167)

Precisely because urban renewal *is* such a visible issue, it has received, therefore, not only empirical investigation, but theoretical scrutiny as well. Before examining these developments, let us outline the impacts of renewal on neighbourhoods.

Renewal as externality

Although the superficial intention of 'renewal' must be ultimately some improvement in the physical fabric of an area, there are various reasons why the process can be intensely threatening and disruptive.

First, the needs of a particular neighbourhood are subsumed within the housing (and other) needs of the urban area as a whole. Mason, for example, writes that:

the fact that the systems of housing allocation do operate on a city-wide basis in both the private and the public sector leads to the possibility of conflict between different areas. The application of similar arguments to planning, environmental health and education severely reduces the extent to which genuine decision-making can be decentralised to sub-area levels. (1977, p. 92)

In other words, there exist three types of tensions in this context: between small areas and the borough or district administration, in relation to the plans that exist for that small area; between the rest of the district or borough, in competition for scarce resources such as improvement area status; and between the district or borough as a whole and the state, in relation to the funding of projects either directly or via the Rate Support Grant (for further discussion on this theme, see Chapter 7).

Secondly, there exist enormous lead times in the implementation of local plans; Mason, for example, notes a gap of 21 years between a locality being scheduled for demolition, and the first survey being undertaken. In such circumstances a self-fulfilling prophecy of decay and non-investment is likely to occur. Inefficiency, changing political circumstances and differing government strategies may all add to this process of delay.

Thirdly, renewal may be prompted not by housing issues (or even as we have hinted already, transport requirements), but by the speculative interests of the private sector. Harvey, for example, has commented extensively on the relationships between the built form of the city and the increasing importance of speculative developments as a means of utilizing the growing amounts of surplus tied up in financial institutions, like pension funds and insurance companies. In such situations old, but sound housing may be removed to make way for offices, leisure facilities, hotels and apartments, which are built and owned privately, but encouraged by public-planning authorities on rating and employment criteria.

Each category of mismanagement and/or poor planning (from the point of view of those displaced) has been well documented. The seminal analysis is that of Jon Davies, based on a small number of streets in West Newcastle called Rye Hill. Although the area was originally scheduled partially for demolition and partly for renovation, little was attempted for over a decade. Home owners were unable to sell their property, due to the uncertainty engendered, whilst the council properties were increasingly filled with so-called problem families. Between 1962 and 1969 only seven properties were modernized, and by that time the area had achieved some notoriety for violence and prostitution. Only prompt action in the 1970s has restored some stability to the neighbourhood. As a comment on this history, Davies observes that 'planning becomes a highly regressive form of indirect taxation, with those who have least suffering most, and those who have a lot being given more' (1972).

Mason, in his study of Manchester and Salford does not subscribe to the 'powerful influence commonly attributed to planners', and emphasizes instead the fact that a local plan is unlikely to emerge

from a consciously followed set of social priorities for the area, but as a result of co-ordination of existing plans that (are) themselves largely determined by generalized clearance and rebuilding policies based on city-wide, departmental goals rather than special integrated policies for particular areas of the city. (1977, p. 78)

In a sense, this simply attributes a negative power of neglect to planning (rather than a positive influence), which overstates the case. The lead time of development has been mentioned, and it must be pointed out that state housing policies have changed several times over the last decade. In many cities it is possible to find adjacent streets of identical properties, some of which have been improved to GIA standard, some of which have been designated as a Housing Action Area, and some that have been cleared, either to be left as vacant land or to be rebuilt upon. Each solution reflects the particular strategy promoted at any one time by the erstwhile Ministry of Housing and Local Government, and latterly the Department of the Environment. In consequence it is more realistic to see local plans as a palimpsest of differing influences over a long period of time.

In contrast, we may examine the 'property machine', a phrase used by Ambrose and Colenutt in their book of the same name

(1975). In two detailed accounts they outline how local authorities of both major parties may promote private redevelopment schemes. In the case of Southwark, for example, residential areas were cleared to make way for office developments at the height of the shortage of office space in London in the early 1970s. As the authors show, the schemes were only of benefit to the local authority in an indirect sense, via the rating system; in direct terms, homes were lost, and replaced by employment opportunities that did not fit into the mould of skills on offer in an inner-city borough. A more insidious process is replacement of local shops by new, larger shopping precincts and parades: 'this suits the planners, developers and shopkeepers, but it is doubtful if local residents, especially the elderly who find it difficult to get about, are satisfied'. (Ambrose and Colenutt, 1975, p. 96)

Protest and community action

Residents cannot live without a home or in a house collapsing about them, although they can live with the noise from an airport or a road for a time. In consequence, planned threats to homes tends to produce speedy and vociferous conflict. This does not imply, however, that this community action, as it has come to be called, is necessarily any more successful than the other cases we have considered. Ambrose and Colenutt identify four 'problems involved in basing a strategy for radical change on community action'. These are that community groups are 'parochial'; that they are 'frequently dedicated to conservation'; that they are transitory; and that a gain in one location will simply 'export the problem to some other area which is "softer" ' (1975, p. 182).

Bassett and Short have also considered these issues, and have produced a typology of community action: 'protest unorganised' (an example of which is Norman Dennis's work on Sunderland, 1972); 'protest blocked' (in which they include Davies's study of Rye Hill) and 'protest partly successful' (1980). The implication of this categorization is again that spatially based protest is frequently unsuccessful, and can never hope to achieve more than a partial victory. A good example of the limited success that can be achieved is provided by Harrop, in a study of a small, recently built council estate in inner Gateshead that was bisected by a flyover section of the A1. Although the estate was unable as a whole to divert the road, residents of several blocks of maisonettes in extremely close proximity to the highway were vociferous enough to have themselves rehoused, and the blocks, less than 10-years old, demolished (Kirby and Harrop, 1976).

This relative failure, in which a draw represents a victory and outright defeat is normal, has caused community action to be roundly criticized, particularly by those with a radical perspective. Cockburn complains bitterly of the way in which community action has been exploited by political parties – from the Liberals to the Communists (CP) – as a virtual alternative to ideological politics at the local scale; this she sees as an obfuscation of the true cleavages in society, which are production and reproduction. (1977, pp. 158–63).

Ambrose and Colenutt express similar sentiments concerning the ideological weakness of community action, although they argue that community groups should link up with mainstream political groups in order to strengthen the latter's position, and to mobilize support for their own causes:

the long-term significance of community action can be judged in terms of its impact on the organisation and composition of political parties, tenants associations and trades unions. Of particular importance is the wider use of the power of trades unions in community action, not only in general support but also in the use of selective industrial action. (1975, p. 183)

In one sense this contradiction illustrates rather well the problems faced by community activists. Every strategy, every response produces both gains and losses; and it is probably true to argue that there is no single course of action which is always correct in every context. Two examples will serve to illustrate this.

The first is the formation of the Battersea Redevelopment Action Group, to contest the redevelopment of an 11-acre site in Wandsworth, on which it was planned to replace an industrial plant with office and luxury housing developments. The group began in 1973, in a history that has obvious parallels with the Southwark case discussed above (SCAT, undated, pp. 31–4). BRAG's strategy was to back up the borough council (which opposed the development at the subsequent public inquiry) by submitting an alternative re-development plan. This involved hiring a solicitor and attending the inquiry on regular basis.

Diametrically opposed histories are recounted by O'Malley in her very readable account of community struggles in Notting Hill. Several of her examples involve direct action: squatting, the invasion of council meetings, the blocking of the M4 Westway and breaking in to private play areas (1977, pp. 75–88). As she notes:

the struggles which involved both centres of the Notting Hill People's

Association were aimed at forcing the local authorities to carry out their planning policies in a socially responsible way, and at forcing them to have policies and spend money in areas which they had never thought about, like play schemes and the development of the space under the motorway. (1977, p. 75)

In both cases then, the strategy was entirely different: it cannot be stated, however, which is a 'normal' or 'correct' approach. The BRAG case could only have been fought in the context of the inquiry, given that the local authority was already in agreement with their stance. If Cockburn's remarks are taken further, it is not easy to fight capital concerns, which may hide behind subsidiary companies based throughout the country or even abroad, except at formal confrontations such as an inquiry.

Conversely, the success of the Notting Hill protests was directly a product of their vocal nature and in direct proportion to the embarrassment caused to the borough council. The organization was minimal, and the success achieved by groups of residents was achieved virtually by individual intervention:

by the end of 1970 only one family was left. The O'Sullivans, a family with 12 children, seemed to have been forgotten. Maggie and Bridie rushed back to the area and gave a seven-days ultimatum to the authorities. If the family was not rehoused by then, they would be forced to consider direct action again. The O'Sullivans were rehoused within the week. (O'Malley, 1977, p. 79)

This is not to suggest that the Notting Hill activists operated in a political vacuum. Their umbrella People's Association possessed links with both the Labour and Communist parties, plus more occasional bursts of solidarity with unions such as those representing the postal workers and building workers. The development of stronger ties was inhibited, however, not by the 'parochialism' of the community, but the nature of the area itself (which lacked large industrial concerns and consequently trade unions) and even the opposition of other political activists, whose 'emphasis on industrial work led them to see many of the activists, who did not work full-time for a wage, as marginals and irrelevant to the real struggle for socialism' (O'Malley, 1977, p. 72). Moreover, the very nature of communities facing housing and renewal problems militates against strong and consistent community action, simply because those being rehoused leave the area:

the issue of the people's need for rehousing was fought and established but just as in Camelford Road, the winning of earlier rehousing meant the destruction of street organisation, so in Walmer Road the winning of the right to rehousing meant the end of street organisation there too. This raises very fundamental questions about the potential of redevelopment areas as a base for class organisation. (O'Malley, 1977, p. 81)

Urban social movements

O'Malleys's observations bring us back to a fundamental issue, namely the extent to which local conflicts (and even the rather wider spatial cleavages identified in Chapter 5) can exist as anything other than small, ephemeral movements that stand distinct from the mainstream political protest. Much British research regards a process of politicization as unlikely: precisely because it is spatially based rather than ideologically based, with the result that existing allegiances are fragmented and undermined: 'community action . . . is often premised on a consensual and harmonious view of community which ignores class differences by emphasising common territorial interests' (Saunders, 1980, p. 128). Consequently, the local nature of the issues, the defensive nature of the conflict and, more importantly, the resilience of those in power, are all seen to conspire to contain the struggles to small areas and limited time periods, until political self-interest transcends the artificial 'spatial community'.

This interpretation of the importance of the material discussed above is a particularly British one: in the Spanish and French contexts a particularly urban sociology looks to conflicts arising from collective consumption – particularly in the field of state housing – for a rejuvenation of radical politics. Research into opposition to urban renewal and similar threats leads Castells to argue that 'urban movements are those which most unify the interests of various classes and strata against the dominant structural logic, and which lead them to confront a state apparatus' (1978, p. 151). Moreover, the emphasis is upon working-class rather than middle-class protest: 'the lower the social strata, of the [social] base, the easier it is for a revolutionary movement to take root politically [*sine qua non* being that it is locally-based] (Castells, 1978, p. 122).

The inapplicability of Castells's arguments to the British case is very well argued by Saunders (1980, pp. 113–36), and there is nothing within the present discussion which could constitute evidence for the emergence of urban social movements. This said, it remains for us to evaluate the importance of the conflicts and cleavages discussed above. Although they may not have provided a springboard for (revolutionary) politicization, it could be argued that a particular political consciousness has developed from the publicity attending opposition to road developments, airport runway extensions, and nuclear power-plant construction. This consciousness manifests itself in the membership of environmental pressure groups and limited electoral support for the Ecological Party. However, as Sandbach notes:

the environmental lobby in Britain (and elsewhere) in its apolitical stand has failed to comprehend the implications of Windscale, nuclear-power policy or motorway policy within any coherent political or sociological perspective. They have been duped into the dominant political ideology of pluralism whereas in fact the pressures to develop nuclear power, waste reprocessing at Windscale and a motorway network are closely related to the nature and demands of modern capitalism. (1980, pp. 124–5)

This is clearly correct, although the criticism is harsh; after all, this critique can self-evidently be applied to virtually every western nation state, and it need not be applied solely to environmental issues. Moreover, it also suggests that existing environmentalism has in the main not emerged directly from spatially based protest. As we have seen, involvement in such conflicts can lead to violence and a certain increase in what we might term radical insight: at the least, it produces a healthy cynicism. There is a gap between these attitudes and the starry-eyed views outlined by Sandbach.

The speculations lead me to suggest that a brand of environmentalism has fostered and encouraged a willingness to at least question the legitimacy of many public actions, and a willingness to learn from the more heated opposition to urban renewal schemes. These spatial cleavages and conflicts have as yet not filtered back to stiffen the environmental lobby; they certainly have not fostered radical social forces. The extent of their potential remains, however, intriguing.

Notes

1 This material is drawn from an analysis by E. Rafferty, (1975) entitled 'The Area 5 experiment: results of a planning decision', University of Newcastle-upon-Tyne, mimeo.
2 The document also included traffic-flow data and an outline of the pros and cons of the four routes.
3 *Yorkshire Post*, 20 July 1977.
4 *Sunday Times*, 12 February 1978.
5 Department of the Environment *Public Inquiries into Road Proposals*, DOE, London, 1974.

References

Adams, J. (1977) 'The breakdown of transport planning,' *New Society* 16 June, 548–60.

Ambrose, P. and Colenutt, R. (1975) *The Property Machine*, Harmondsworth, Penguin Books.

Bassett, K. and Short, J. R. (1980) *Housing and Residential Structure*, London, Routledge & Kegan Paul.

Castells, M. (1978) *City, Class and Power*, London, Macmillan.

Cockburn, C. (1977) *The Local State*, London, Pluto Press.

Cox, K. R. (1979) *Location and Public Problems*, Oxford, Blackwell.

Davies, J. G. (1972) *The Evangelistic Bureaucrat*, London, Tavistock.

Department of Transport, UK (1978) *Report on the Review of Highway Inquiry Procedures*, Cmnd. 7133, London, HMSO.

Dunleavy, P. (1980) *Urban Political Analysis*, London, Macmillan.

Goddard, J. B. and Kirby, A. M. (1976) 'An introduction to factor analysis', *CATMOG*, 7, Norwich, Geo-Abstracts.

Grant, J. (1977) *The Politics of Transport Planning*, London, Earth Resources Research.

Humphrey, C. R., Bradshaw, D. A. and Krout, A. (1978) 'The process of adaptation among suburban highway neighbours', *Sociology and Social Research*, 62(2), 246–66.

Kirby, A. M. and Harrop, K. J. (1976) 'Some preliminary observations concerning the problems of quantitative explanation and social geographic research', *NEAS Working Paper 29*, Department of Geography, University of Durham.

Lassiere, A. (1976) 'The environmental evaluation of transport plans', *Research Report 8*, London, Department of the Environment.

Leitch, G. (1977) *Report of the Advisory Committee on Trunk Road Assessment*, London, HMSO.

O'Malley, J. (1977) *The Politics of Community Action*, Nottingham, Spokesman Books.

Malpass, P. (1979) 'The politics of participation', *Architects' Journal*, 16 May, 1012–13.

Mason, T. (1977) 'Inner city housing and urban renewal policy', *CES Research Series 23*, London, Centre for Environmental Studies.

Mumphrey, A. J. and Fromherz, C. B. (1978) 'Citizens, politicians and decision makers: a Helix game for transportation planning', *Geoforum 9*, 279–91.

Plowden, W. (1973) *The Motor Car and Politics in Britain*, Harmondsworth, Penguin Books.

SCAT (undated) *Public Inquiries Action Guide*, London, Shelter Community Action Team.

Sandbach, F. (1980) *Environment, Ideology and Policy*, Oxford, Blackwell.

Saunders, P. (1980) *Urban Politics*, Harmondsworth, Penguin Books.

Smith, D. M. (1977) *Human Geography – a Welfare Approach*, London, Arnold.

Twinn, I. (1978) 'Road Planning: a critical analysis of public involvement', unpublished PhD thesis, Department of Geography, University of Reading.

Wheeler, J. O. (1976) 'Locational dimensions of urban highway impact: an empirical analysis', *Geografiska Annaler B*, 58(2), 67–78.

Part IV

Change

This concluding section is essentially speculative in nature, although the two chapters presented are very different in intention. Chapter 7 follows on from the material discussed in Part II, and again takes up the theme that spatial organization may result in forms of deprivation. Here, however, the emphasis is upon types of practical remedy, in particular those which improve the 'quality of life' by using spatial policies: the Rate Support Grant is a central example.

Chapter 8 (and the final discussion within the book) moves away from these specific questions in order to evaluate the arguments presented here as a whole. Inevitably, an author's ideas are in a state of flux, and any long work must contain in some sections material that is newer, and probably more rigorous than in others. Equally, no argument can make every qualification and tie up every loose end as it proceeds. Chapter 8, therefore, is an attempt to point to the weaknesses of the argument, the under-statements and the likely criticism; not necessarily to forestall them, but rather to highlight the overall thesis.

.

7 Spatial change and social change

There is a limit to which you can influence things which are inherently beyond your control.
> (Will Cuppy. *The Decline and Fall of Practically Everybody*)

Spatial change and social change

I should not now need to reiterate that social affairs proceed in three dimensions, and that changes in the spatial economy and the inequitable distribution of phenomena between locations have important (often deleterious) impacts upon populations. Increasingly, this message is being learnt, and political remedies that have implicit spatial form are the result: the Assisted Areas legislation and the Inner Areas Partnerships are British examples that have already been well discussed elsewhere.

However, this is not the end of the story. Instead of simply seeing these types of planning exercise as renovative, we can identify several contexts within which spatial change may inaugurate social change. Those who are wary of 'fetishism' in relation to space (and there are many: see Chapter 8) will regard what are to follow as examples of 'bleeding-heart liberalism' – small-scale tinkering with man's 'striving to be unequal', as David Smith describes it. In part this is true: the social changes envisaged remain within a particular social and political context, which is itself far beyond improvement by even the most radical 'reforms'. Indeed, it is this very permanence of the existing social system that causes me to emphasize the importance of spatial changes, simply because these are one of the few directions from which the present social system can be in any way be improved.

Once more, this statement needs immediate qualification: too much can be expected from physical solutions to social issues. The utopian planners of the nineteenth century envisaged that spatial engineering could recreate urban society:

According to this view, problems of social malaise in the city will be met by building a new environment to replace the old – whereupon poor health, inadequate education, badly balanced diets, marital discord and juvenile delinquency will all go away. (Hall, 1975, p. 80)

Physical planners have begun to slough off this legacy of physical determinism, although its mantle now seems to have fallen in turn upon architects. Commentaries upon the role of this profession are few, although it is clear that, in Rubin's words, 'aesthetic ideology remains a potent vehicle for the perpetuation of urban, economic and social inequalities' (1979, p. 361). Attempts to apply the bold physical concepts of design and layout produced by architects like Le Corbusier, have frequently resulted in increased social and personal tensions (see, for example, Ardagh, 1977) and a shotgun marriage between the weakest in the housing market and the worst design; Taylor, for example, relates the history of an apartment block in Killingworth Township (North Tyneside) which was grandly designed as a 'medieval city', but which has locally earned the name of Alcatraz (1979b).

Utopia is not achieved through bricks and mortar, therefore. Nevertheless, physical responses can play a modest role. The corollary of a relationship between design and unhappiness is that physical change may improve matters; one of the practical remedies proposed within the Lambeth Inner Area Study report was that redesigning the spatial layout of many council estates could reduce vandalism – the removal of 'neutral space' would avoid areas in which destruction can proceed unchecked (1977). The rest of this discussion is pitched at this level, seeing spatial changes not as central intrusions into a particular situation, but rather as major aspects of detail. We begin with the redistricting of school catchment areas.

Segregation and education

The spatial organization of education was considered in Chapter 3, and it was suggested that the social segregation that exists within most cities can, when allied to some organizational system such as feeders or catchments, produce schools with significantly different mixes of pupils. This is not unknown to education officers, who periodically redistrict the boundaries. An example of this is discussed in some detail in Kirby (1979b), in relation to a redistricting undertaken by the education department in Newcastle upon Tyne in 1977. In that exercise specific social goals were outlined – namely, the creation of neighbourhood schools and an improvement in the social mix within the comprehensives – and a particular spatial organization of the school feeder system was designed.

This example is revived, not in order to remind us how a particular goal may be achieved, but rather to illustrate what can go wrong between intention and solution. Those familiar with educational issues will have already noted that local, community schools must perforce draw upon a relatively small, local catchment area or set of feeder schools. However, this brings us back to the issue of socio-spatial segregation: in other words, local schools will tend to have a very one-dimensional mix of children, who all come from similar backgrounds in terms of housing class, income or whatever. In consequence, the two stated social goals are mutually exclusive, and in fact, the redistricting in question produced an increase in polarization between comprehensives.

Particular social goals require detailed specification, therefore, as has been illustrated in the United States. There, tension has arisen in relation to racial segregation, which was traditionally maintained through the school system. Following the Supreme Court's ruling that such practices were unconstitutional, counties were responsible for redistricting exercises which ensured that the racial balance within the community at large was expressed within each of its schools, be it a white/black ratio of 95/5 or 50/50.

In order to achieve such a balance, relatively sophisticated statistical devices may be employed. Various indices have been designed, which measure the degree of segregation extant within a particular school, and the proportion of children of a particular race that would have to be moved in order to achieve a satisfactory balance (Zoloth, 1976). When this is known, practical steps to redesign the school boundaries can be undertaken, using powerful mathematical tools such as linear programming. A good example of this is given by Maxwell, in a study of changes to the black and white catchment areas in Athens, Georgia. This is reproduced, alongside an outline of the procedures involved, in Taylor (1977), and it can be seen that to achieve a suitable racial balance, highly complicated redistricting exercises are required.[1]

This type of example provides probably the clearest justification available for the development of spatial analysis (should any still be required), although it is by no means an isolated example; Haynes and Bentham, for example, provide a detailed account of the spatial changes that have been wrought by the National Health Service, particularly in rural areas, in attempts to provide acceptable (and accessible) levels of medical cover (1979).

Wider issues (1): the Rate Support Grant (RSG)

Although decision-makers may be unwilling or unable to implement such measures, there is a fairly clear relationship between the social goals outlined above and the spatial changes necessary to bring them about. We may now turn to a far more complex issue, namely the operation of the Rate Support Grant. It is complex in different ways (the calibration of the needs grant is in itself, as we shall see, worthy of the description), but principally in so far as it is *not* intended as a means of redistribution between different local authorities, but none the less functions in such a manner.

The RSG is the principal intervention by the state in the affairs of local government, namely as part-financier of local authority activity. This involvement has steadily increased year by year, until it now accounts for approximately half of all local government finance. Prior to changes in 1981 money was allocated in three ways, as Table 7.1 outlines.

Table 7.1 The components of the Rate Support Grant, pre-1981

Component	Aim	Amount (c.f. total RSG)	Calibration
Needs element	To take into account local needs, at-risk populations	Approx. two-thirds	Regression formula
Resources element	To compensate for low rateable values and a reduced rate income	Approx. one-quarter	Compensation to reach some standard value of rate income per head
Domestic element	To provide rate relief for domestic rate-payers where commercial enterprise lacking	Approx. 10 per cent	Compensation of 18.5 per cent of total rate bill to each domestic ratepayer

The resources and the domestic grant are relatively straight-forward and operated as follows. The resources element was introduced in 1974, and was intended to compensate declining localities where a depleted tax base meant that the amounts raised by local rates were low: the compensation was designed to bring authorities up to the average level nationwide. After 1975/6, all metropolitan districts were in receipt of the resources element

(Jackman and Sellars, 1977). The domestic element had a similar compensatory intention, although in this instance the aim was to bring about rate relief for householders *vis-à-vis* commercial-property holders. In this case, therefore, all properties were initially rated at a unitary level, and an overall rate income computed for each authority. The contribution for each domestic ratepayer was then scaled down by 18.5p in the £, and it is this shortfall that was, in aggregate, met by the RSG.

The needs element is a good deal more complex, both in its formulation and its calibration. As Harrison and Jackman observe,

no government has defined the purpose of the needs grant . . . but [the Labour] Government did refer to the amount needed to 'provide a level of service comparable with other similar authorities'. We take it that a local authority's 'need', for needs grant purposes, is the amount it 'needs' to spend in order to provide some common standard of services. (1978, p. 22)

This amount will vary for several reasons, which include variations in the client populations (concentrations of the young or the elderly), geographical differences (variation in population density, road mileage, age of dwellings) and monetary pressures (notably different labour and building costs in different areas).

The needs element aimed to counteract these variations 'in order that ratepayers should not have to face differences in their bills purely because of where they live'. This is an equitable intention, but one that gives little guidance as to how equalization should occur. Until 1974 needs were determined by recourse to a fixed formula that took into account numbers of school children, road mileage and so on. After that date, however, a regression approach was adopted:[2] 'the regression analysis seeks to identify characteristics of local authorities which are associated with greater spending and bases the grant on these characteristics' (Jackman and Sellars, 1977, p. 20). There is thus an analytical element involved, as attempts are made to isolate these sectors on which authorities have to (or simply want to) spend heavily. For example, if all authorities spending heavily were shown to have large concentrations of single-parent families, we might expect the needs element to provide cash in proportion to single-parent numbers.

The resolution of the regression formula, by which significant variables of need are chosen, was complicated by the fact that there are several of these key variables. The latter are, moreover, often closely interrelated, which can lead to statistical errors (for example, at-risk groups are often associated with poor housing for financial reasons; placing both variables in a regression equation as indicators of need may be akin, however, to adding the same

Table 7.2 Different independent variables utilized in needs element of the RSG, 1974–8

Variable (measured for each local authority)	Year			
	1974/5	1975/6	1976/7	1977/8
Regional weighting	√	√	√	
Labour-cost differential				√
Metropolitan high-cost areas	*	*	*	*
Population of area	√	√	√	√
Population decline	√	√	√	√
Low density	√	√	√	√
High density		√	√	√
Personal social-service units	√	√	√	
Education units	√	√	√	√
Children under 5 years	*	*	*	*
Children under 15 years	*	*	*	*
Pensionable age		√	√	√
Unemployed				√
Lone parents			√	√
Shared households				√
Lacking basic amenities			√	√
Room density			√	√
Housing starts			√	√
Road mileage	*	*	*	*
*=in use pre-1974				

Source: adapted from Jackman and Sellars (1977) Figure 1.

variable twice). Table 7.2 indicates the diversity of variables used in RSG calculation (it is to be noted that these variables did not necessarily apply in the calculation of London's need element: see Jackman, 1979).

Problems such as multi-collinearity (as these interrelationships are termed) have consistently undermined the RSG calculations, and this fact alone accounts for the introduction and disappearance of different variables over time. This problem, allied with the 'age' of 1971 census data and the non-inclusion of political variables (such as Labour or Conservative control in an authority), suggests that regression equations may simply describe the types of authority that spend heavily – they may not explain (or predict) that expenditure. This is not an academic point: it may account for the major shifts in the needs element that have occurred since 1974:

the formula is changing to provide more grant for the cities because they are spending more . . . needs have been growing more rapidly in the cities due to change in the types of services provided and suchlike . . . but unless there is some independent measure of needs, this argument becomes circular: if an increase in expenditure is taken as evidence of increased needs, we are explaining an increase in expenditure by an increase in expenditure. (Jackman and Sellars, 1977, p. 30.)

According to this type of interpretation, the RSG may simply be picking out independent variables that describe high-spenders, rather than the needy:

the grant mechanism could 'snowball', an increase in grant leading to an increase in expenditure, leading to a further entitlement, a further increase in expenditure, and so on. A whole group of authorities with similar characteristics can, by increasing their spending, increase the weight given to these characteristics in the regression equation, and hence influence the grant distribution in their favour. (Jackman and Sellars, 1977, p. 30.)

This discussion leaves us with two possible interpretations: either that the regression formulas are successful in tracking down need, or that high-spenders are in practice simply attracting a larger needs element. Either way the RSG is clearly an important tool for distributing public funds to different spatial units. This point is underlined by Jackman and Sellars, who compare the changes 1974–1978 with what would have happened had the fixed formula remained. Their conclusions are straightforward: the shire counties lost expenditure, and the metropolitan counties made gains, although the bulk of the change was in the form of massive gains by the London boroughs.

Table 7.3 Gains and losses due to changes in the RSG 1974–8: figures in percentage points; extremes and medians only

Shire counties		Metropolitan districts		London boroughs	
Surrey	−114.41	Solihull	−6.92	Havering	+50.91
Herts	−59.49	Leeds	+41.17	Croydon	+85.09
Cleveland	−0.64	Tameside	+63.31	Haringey	+131.30

Source: adapted from Table 2, Jackman and Sellars (1977, p. 27).

The interesting point to note concerning the RSG is that this change in its balance represents simply one of many possible recalibrations of the regression formula; different independent variables would have altered the direction of funds markedly. Such changes could take into account political complexion (although this is clearly involved in the shire–metropolitan shift outlined already); nationality (Scotland, Wales, N. Ireland); Assisted Area status;

immigrant concentrations and so on. It needs to be restated at this point that, at present, the RSG is not intended as an explicit redistributive agent. The direction of cash to metropolitan areas, which accorded with the 1974–9 Labour Government's commitment to the inner areas of these large cities, was unplanned: 'it is purely fortuitous that the regression method has led to a pattern of grant reallocation that accords with this Government's political preferences' (Harrison and Jackman, 1978, p. 26).

None the less, a redesign is possible, and has been investigated by the Centre for Environmental Studies team. They have, for example, outlined a series of different funding systems; these include a pure centralist model, some form of consensus agreement on levels of service provision, a setting of minimum standards, and finally a pure localist system in which local authorities all receive a similar amount, which they have discretion to spend as they like (Harrison and Jackman, 1978; Harrison, Jackman and Papadachi; 1979). They show that virtually any alternative to the regression approach would have reduced the variations between authorities that existed in terms of grant received; the range that extended from £40 per head through to £290 per head in 1979 would shrink to a range of £40–£140 if some form of *per capita* assessment was emphasized, with the varying numbers requiring educational or social-service spending being identified. Behind these aggregate differences would also be, of course, a whole series of individual changes, with some authorities suffering radical alterations to their relative 'income'. Once again, these can be summarized in terms of the shire counties, the metropolitan districts and the London boroughs, as Table 7.4 shows:

Table 7.4 indicates clearly that London is particularly vulnerable to any changes in the RSG; as it has high concentrations of 'need', it naturally receives a large amount in terms of the needs element. (London in fact does so well that some money is 'clawed-back' and redistributed amongst other recipients; this is done because of the capital's high rateable values, particularly in relation to commercial property, which allow most boroughs to raise high rate incomes from relatively low rate poundage (see, for example, Jackman, 1979).

By way of summary we can see that between 1974 and 1979 the RSG achieved a certain equilibrium, in which certain types of local authority were heavily funded and others lightly funded. This does not necessarily mean that the inhabitants of some authorities had low rate bills – often quite the opposite. It does point, however, to areas in which municipal involvement, in terms of the provision of

Table 7.4 Rank changes in needs grant per head (£) under 1979 and alternative (per capita) systems

1979 regression expenditure		Alternative (per capita)	
Authority	Rank	Authority	Rank
Westminster	1	South Tyneside	1
Islington	2	Trafford	2
Tower Hamlets	3	Knowsley	3
Southwark	4	Manchester	4
Hackney	5	Brent	5
Camden	6	Cleveland	6
Hammersmith	7	Doncaster	7 =
Kensington and Chelsea	8	Solihull	7 =
Lambeth	9	Bolton	9
Lewisham	10	Liverpool	10

Source: adapted from Harrison, Jackman and Papadachi (1979); for detailed discussion of the *per capita* scheme see this paper, p. 30.

services such as education, was at a maximum. As a means of intervention the needs element was thus particularly powerful; it could, following legislative change, have been used as a means of bringing local authority standards of provision up to some minimum level, by funding accordingly. Conversely, it could have been used as an explicitly spatial tool in the manner outlined above, perhaps in relation to regional policy, or some new goal such as a reduction in inner-city expenditure and an increase in rural investment.

Following the introduction of a Tory administration in 1979, major changes have again occurred in the organization of the RSG. Most importantly, the needs and resources elements have disappeared, with the introduction of a new, block-grant system; this may be outlined as follows:

$$G = RE - [SRP(RE, SE) \times GRV \times \beta]$$

where G is the block grant;
RE is relevant expenditure, as defined by central government;
SRP is standard rate poundage;
SE is standard expenditure;
GRV is gross rateable value;
β is some (changeable) constant.

Simply, the central administration is now able to define arbitrarily a local authority's needs (SE); when local resources are taken into account (GRV), a prediction of the local rating level (SRP) can

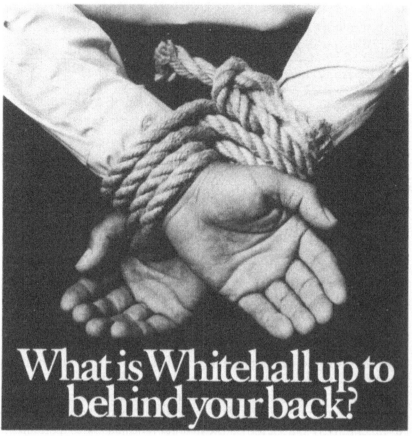

What is Whitehall up to behind your back?

Figure 7.1   These posters, issued in October 1981, illustrate the response of the Association of Metropolitan Authorities (e.g. West Midlands, Merseyside, Tyne and Wear, Manchester, Yorkshire) to Environment Secretary Heseltine's plans to control local spending: this the Association described as 'a shift of power from democratically elected councils to the centre'.

As various commentators have noted, however, many communities both in Britain and the USA have successfully opposed particular levels of spending: California's Proposition 13 is perhaps the best known. This attitude is reflected in the satirical magazine *Private Eye*; beneath the caption 'Why is Whitehall trying to stop us stealing your money?', an inversion of the second poster's intentions. The 'new' poster continued:

'Know what [Heseltine's] trying to do? He's saying that councils like the GLC would have to hold a referendum every time they want to slap on a supplementary rate. Its bloody disgusting. Would you believe it? Heseltine is actually going to force us to consult you before the rate demands go out.

Its downright undemocratic. What the hell do a lot of miserable ignorant ratepayers know about the financing of local government?

The fact is that we in local government have an inalienable right to spend your money just as we like – on essential services such as putting full-page advertisements like this in every newspaper in the land.'

be made; this is then subtracted from the overall estimate of expenditure (RE), and the residual is provided via the RSG. A novelty in this schema is that 'over-spending' is punished quite harshly by the withdrawal of grant.

The redistributive consequences of the block grant are subsidiary – as far as central government is at present concerned – to the overall limitation of local spending (see, for a further discussion of this interpretation, Kirby, 1982a). None the less, such consequences have occurred, and the results have been predictable;

Watch out. Whitehall has plans for your local elections.

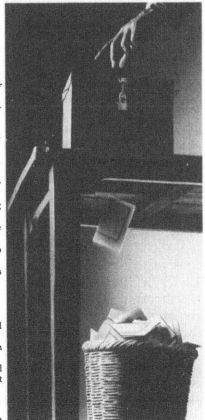

There's some very worrying legislation about to creep in and out of Parliament.

The idea is to take away your Local Authority's power to levy rates.

If you hate rates (and who doesn't), you could be fooled into believing it's good news.

That's what Whitehall is relying on.

But think. Without money your council is also without power.

It can't make decisions. It can't go against Whitehall. Even if you want it to on certain issues.

That's the value of your local council.

It can check excessive control of local affairs by any Government.

Remember, after an election the Government does not have to be nice for five years.

When you come to us with your problems our hands will be tied.

We'll both come up against this innocent looking law. And like all laws, just try arguing with it.

It won't matter if your local councillor agrees the roads are bad (he lives there too).

It won't matter if classes at the local school are too big (he'll probably have children there).

It won't matter if there's no room at the old people's home for our senior citizens.

There will be no point in appealing to us.

In fact there will be no real point in electing councillors at all.

As things are, our doors are open. Whitehall's will stay closed.

Governments ask you to give them your vote when it suits them.

Make no mistake. With this legislation, as far as local elections are concerned, they might as well take your right to vote away.

KEEP IT local

formerly high-spending authorities (notably the metropolitan districts and London boroughs) have lost their relative advantage. Nor has the changing relationship between central and local government been lost on the local outposts of the state; as Figure 7.1 indicates, a vigorous campaign has been launched to bring this to the attention of local voters, although many doubtless would agree with the sentiments proposed within the caption, which neatly encapsulates the growing response of many communities to what are perceived to be unacceptably high levels of public spending.

Public finance, as this section has hinted, is complex in both a technical and a political sense. Through a medium like the RSG, the state is in a position to dictate levels of spending within local units, and to direct central spending where it will. The sheer scale of funds involved in this process argues that this (spatial) control is an enormous one, with important redistributive possibilities.

Wider issues (2): electoral reform

In Chapter 4 we examined the spatial bases of the electoral system, and it was clear that alterations to constituencies, in terms of both their size and their boundaries, can result in very different political outcomes. This section takes this idea a little further, and considers the political implications of various types of electoral reform.[3]

Initially, the distinction between reform and gerrymandering should be made. There are several examples of the latter, where societies (or more correctly, those in a position of power) have determined that certain types of political party should not prosper, and have achieved this end via spatial rather than legislative means. The most extreme example is that of France, which has frequently changed the details of its electoral system with the intention of limiting the left-wing vote. In 1927, for example, legislation dictated that urban areas should exist as separate constituencies, with the result that some Communist candidates were successful. In retaliation redistricting has since tended to swamp urban seats with blocs of conservative rural votes; a method, according to Gudgin and Taylor, which was first developed by Louis Napoleon in the nineteenth century (1979). Since the First World War France has experienced five changes of electoral system, although the most contrived solution was attempted in 1951. In the election of that year a dual system was in operation: Paris elected its members on a proportional-representation (list) basis, whilst the rest of the country voted in multi-member constituencies, returning deputies elected on a simple plurality basis. This meant that there was a maximum anti-Left bias at work. In Paris socialist and communist strongholds were diluted by the PR system, whilst in the rest of the country the opposite was the case; although scattered left-wing pockets of support would have counted under proportional representation, they were wasted in the plurality elections.

Such extreme examples are rare, and also unnecessary. Major changes in the seat-vote relationship can be achieved by fairly simple means, such as, for example, increasing or decreasing the size of constituencies. As we saw above, political allegiances may be place-specific; consequently, only if the support and the constituencies coincide will a party be successful. For this reason it is, for example, more likely that the National Front will first achieve success in local politics than in parliamentary affairs. Both major

Table 7.5 Westminster and European Assembly election results, 1979

Party	Westminster		Electoral bias (%)
	Votes (%)	Seats (%)	
Conservatives	43.9	53.3	+9.4
Labour	36.9	42.2	+5.3
Liberals	13.8	1.7	−12.1
	Strasbourg		
Conservatives	50.6	76.9	+26.3
Labour	33.0	21.7	−11.3
Liberals	13.1	0.0	−13.1

Source: calculated by author.

British parties are also affected by their concentration of support, and in consequence, 'waste' a proportion of their votes; in 1979 the Conservatives, for example, 'wasted' 1.13 per cent of their share of the two-party vote (Gudgin and Taylor, 1980, p. 521). If constituencies are increased in size, this process becomes even more important, as the following example indicates.

Table 7.5 outlines the electoral bias (i.e. the imbalance between votes cast and seats gained) that obtained in the May 1979 General Election. As we might expect, the Conservatives have the largest positive bias, and the Liberals a large negative bias. Labour, despite losing the election, still picks up more seats than its one-third share of the vote suggests. Within the table comparison can also be made with the results of the June 1979 European Assembly (Strasbourg) elections, for which 78 large constituencies were used (rather than the usual 635 for Westminster elections), the number of seats being dictated by Britain's population *vis-à-vis* the other EEC countries (Taylor and Johnston, 1979, pp. 362–8).

The timing of the elections makes them strictly comparable, in so far as major demographic changes and other 'noise' variables can be excluded; the only factor that differed between the two contests was the turnout, which was in June below half that for the General Election. This may in part be attributed to Labour's equivocal stance on Europe, but there is little evidence that turnout varied greatly between seats won by Labour or Conservatives. When we examine the EEC results, we can see that the bias is far greater than in the General Election. Whilst the Conservative vote increases only slightly, the proportion of seats gained rises dramatically; thus

Figure 7.2 European constituencies, London, May 1979; named are the North-Eastern and South Inner seats, both won by Labour. The map also shows the Westminster seats, identified by the party winning the constituency (C = Conservative; L = Labour).

It can be seen that there is widespread support for both Conservative and Labour parties; however, only two of the larger Strasbourg seats were won by Labour. Although this may partly be attributed to Labour's ambivalent position *vis-à-vis* the EEC in 1979, it is also a function of the districting used to create the seats. Different configurations (or, of course, a form of proportional representation) would yield different electoral outcomes.

with half the votes they capture three-quarters of the seats. Labour in this instance has a large negative bias, which is almost as large as that suffered by the Liberals.

We can begin to explore the cause of Labour's demise with reference to Figure 7.2; which shows the ten constituencies used in the Strasbourg elections (and the ninety-two parliamentary seats) in Greater London. As Table 7.6 indicates, London's results were in line with those of the country as a whole. Of the ten European Assembly constituencies, only two were won by Labour: these were the North-East and the South Inner seats (see Figure 7.2). This

Table 7.6 Westminster and European Assembly election results. Greater London 1979

Party	Westminster		
	Votes (%)	*Seats* (%)	*Electoral bias* (%)
Conservatives	46.5	53.2	+6.7
Labour	39.9	46.7	+6.8
Liberals	11.8	0.0	−11.8
	Strasbourg		
Conservatives	51.6	80.0	+28.4
Labour	37.2	20.0	−17.2
Liberals	10.3	0.0	−10.3

Source: calculated by author.

relative failure is a result of Labour's concentration of voters in the inner areas (particularly Lambeth, Lewisham, Newham, Southwark and Tower Hamlets). In these boroughs, Labour picks up over a dozen parliamentary seats, but at this scale the degree of solidarity merely constitutes an overconcentration of left-wing support and in consequence a large number of wasted votes. Counter-intuitively, Conservative support is distributed far more widely, and the party picks up votes even in the inner ring. This is demonstrated by its success in London North (adjacent to London North-East), which contains a majority of Labour parliamentary seats, but in addition a sizeable minority of Tory voters. When the latter are coupled with more traditional areas of Conservative voting on the periphery, a narrow Tory majority results.

It should, of course, be pointed out that the imbalance of gains and losses in these European Assembly elections in London need not have involved such a large margin. The size of the constituencies, *vis-à-vis* the concentration of left-wing votes, was against the Labour party, but the result could be changed by a redistricting exercise. Drawing on the information in Chapter 4, we can predict that the existing boundaries are only one possible solution: simulations in fact suggest that 7252 exist. As Table 7.7 indicates, a majority of the possibilities could involve Labour successes. An alternative approach which favours Labour would be to split up the

Table 7.7 A simulation of Labour gains in European elections in London, assuming 1974 (October) voting figures

	Seats to Labour						
	1	*2*	*3*	*4*	*5*	*6*	*7*
North of the Thames		31	86	69	10		
South of the Thames	18	19					
Total			558	2137	2876	1491	190

Source: Johnston (1978, p. 23)

cores of support that exist in the central areas, and to range them against the bases of Conservative voting out on the periphery. A solution with constituencies fanning out like the spokes of a wheel would be far more likely to reduce the 1979 electoral bias, although minor boundary changes within each large seat can produce very different results. Figure 7.3, for example, shows various partitions in the two European constituencies in North-East London, and whereas the existing districting and the first repartition produce one seat each for the Conservatives and Labour, the second attempt produces two victories for Labour (Kirby, 1982b). In due course, however, the existing boundaries may be used as the basis for proportional-representation elections to the European Parliament and as a result the existing bias would be reduced without a re-districting being necessary.

Proportional representation

The mention of proportional representation brings us to the central issue of electoral reform: as Johnston observes, 'a majority way of altering the outcome of a contest is to change the rules' (1979b, p. 184). Various examples of non-plurality systems exist, and although each is designed to produce proportionality, different results would obtain in each case. For example, the method used in Australia is the alternative-vote (AV) system, in which the spatial organization

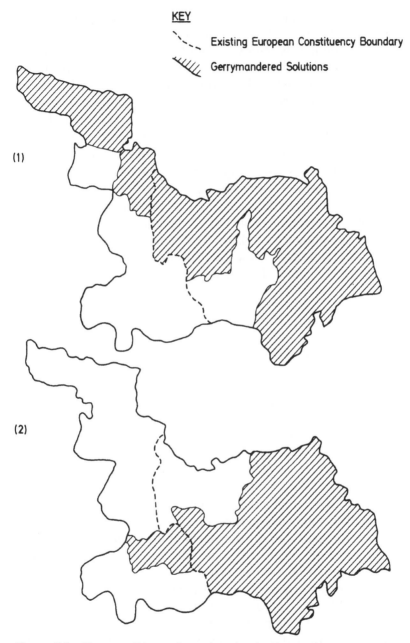

KEY

‑ ‑ ‑ ‑   Existing European Constituency Boundary

▨▨▨   Gerrymandered Solutions

(1)

(2)

Figure 7.3   Two possible configurations for the European election seats,
London North-East and London East, including the existing boundary

The first map shows a districting solution which, on the basis of votes cast in the
Westminster elections, would have produced a result of one seat each for the
Conservatives and Labour. The second map represents a solution which gives both
seats to Labour.

of constituencies is as in Britain, but voters rank all the candidates. Where there is no clear overall majority for a candidate, the lowest-ranking opponent is removed, and his or her second choices are reassigned to the other candidates. This process is repeated until a majority is achieved (Taylor and Johnston, 1979, pp. 419–20). The AV system produces proportionality within each constituency, but would do little for minority parties nationally. In order to benefit the latter, changes in the organization of elections have to be coupled with the spatial reorganization of constituencies.

In the British case two proportionality systems are thought to be suitable alternatives to the existing organization, and both involve radical alterations to the spatial basis of constituencies.[4] The first is the Single Transferable Vote (STV) method, which is used in multi-member constituencies in Eire. STV requires the voter to rank all his or her preferences: these are then allocated to the candidates in the following manner. First, a quota must be met, normally in the form:

$$\frac{\text{Total number of first preference votes}}{\text{Number of seats in the constituency} + 1} + 1$$

If a candidate achieves the quota, his or her 'excess votes' are then reallocated amongst the other candidates. By reallocating votes from the most successful and the least successful candidates, the available seats are ultimately filled (Taylor and Johnston, 1979, pp. 57–9). Despite this complexity, the degree of proportionality achieved is very much dependent upon the size of the constituencies employed. To take a simple example, proportional representation cannot be achieved in a three-member constituency in which there are two evenly matched parties; in such a case the result will always be 2:1 to one party, to the detriment of the other: (a four-member constituency of course overcomes this problem). Similarly, very large constituencies are necessary to aid very small parties. In past instances the Liberals, with approximately 11 per cent of the British vote, would have captured a seat only in nine- or ten-member constituencies. Gudgin and Taylor show, for example, that the Liberal share of seats in the October 1974 General Election would have risen (under STV) to 64, if three- and four-member con-stituencies had been employed (1979); if, however, larger constituencies are used in the simulation, the Liberal share of the seats increases to 102, or 16.4 per cent of the total. Interestingly, it is for this reason that the small Irish Labour Party still suffers a negative electoral bias, as most constituencies in Eire have only three members.

The second system to receive some support is the German Bundesrepublik's Additional Member System (AMS), which employs a mixture of electoral practices; half the seats are elected on a plurality basis, and half are allocated afterwards in order to achieve proportionality; in the latter process regional voting patterns are taken into account. In Germany all three major parties enjoy positive bias, which is achieved at the expense of the very small parties, which do not receive any 'top-up' seats unless they capture at least 5 per cent of the vote or three constituencies. As far as the German elections are concerned, proportionality is achieved between the major parties, but predictions for Britain are not as optimistic. Gudgin and Taylor show that if regional support is taken into account at the top-up stage (as in the German Länder), the Liberals would fare badly, as they lack particular regional concentrations of support. Only if topping-up is applied at the *national* scale (England, Wales, Scotland, Northern Ireland) do the Liberals fare as well as the Nationalist parties (Gudgin and Taylor, 1979).[5]

To summarize, therefore, it is clear that the key to electoral reform is not the form of the electoral system employed, but the spatial basis upon which it depends. As we have seen, even STV can be manipulated (as in Eire) to produce a particular, consistent bias, whilst AV is only capable of achieving proportionality for the voter, not the eradication of bias. AMS is the least sensitive to the spatial element, but even in this case the degree of bias depends very much upon the scale at which topping-up occurs. Once again we return to the issue that was emphasized in Chapter 4, namely the mismatch between ideological political parties and the spatial basis of electoral support; perhaps if a regionally based AMS system were to be adopted, then the trend towards 'Thatcherland' might become a reality, with all that this implies for the future development of political cleavages in this country.

Conclusions

This chapter has tried to outline some of the ways in which spatial engineering could bring about changes in social behaviour. The three examples discussed are very different in terms of their contexts; one relates to local authority activity, one to policy decisions by central government, and one to electoral change that would require major legislative debate. Equally, each would have very different long-term implications; a change in the effectiveness of the educational system, a redirection of funds from cities to counties or from one region to another, and the alteration of political realities by increasing the power of parties like the Liberals – all these would have obvious and unlooked-for results.

I suggested at the beginning of this chapter that these changes could be readily achieved, and this requires some clarification. All these alterations to the spatial organization of society could be easily organized; this does not imply that they figure strongly in any political agenda. There is evidence that the changes in the RSG and local authority organization in Britain post-1974, for example, had far more to do with centre–local relations than the needs of local government (Kirby, 1982a; Taylor, 1982). Similarly, it must be a supreme irony that the German electoral system was designed by the western allied occupation after the Second World War (in order to restrict any possibility of communist or neo-nazi electoral growth), and yet proportional representation has always been decried in the UK itself.

In the final analysis, space is manipulated far more usually for personal gain, be it by local authorities manipulating their boundaries to include or exclude particular phenomena, or by communities using space to maintain their fiscal, political and jurisdictional independence (Cox, 1979). These examples show the possibilities, if not the most desirable applications of spatial manipulation.

Notes

1 Taylor's book also provides, alongside the relevant calculations, a useful discussion of various examples of spatial reorganization in the fields of education and health care.
2 Readers unsure about the technicalities of regression are again directed to Taylor (1977), or in this context, Jackman and Sellars (1977). A fuller discussion of the RSG in the context of public finance is also provided by Bennett (1980).
3 The notion of electoral reform is sometimes confused, in that it may be taken to mean the removal of individual bias (wasted votes) or the problems facing parties such as the Liberals. Taylor and Johnston argue that proportionality should remove both problems (1979, p. 431).
4 The Israeli system of 'at-large' elections has never been suggested as a serious alternative in Britain.
5 These simulations, of course, pre-date the Liberal – SDP alliance of 1981.

References

Ardagh, J. (1977) *New France: a Society in Transition 1945–77*, Harmondsworth, Penguin.

Bennett, R. J. (1980) *The Geography of Public Finance*, London, Methuen.

Cox, K. R. (1979) *Location and Public Problems*, Oxford, Blackwell.

Gudgin, G. and Taylor, P. J. (1979) *Seats, Votes and the Spatial Organisation of Elections*, London, Pion.

Gudgin, G. and Taylor, P. J. (1980) 'The decomposition of electoral bias in a plurality election', *British Journal of Political Science*, X, 515–21.

Hall, P. G. (1975) *Urban and Regional Planning*, Harmondsworth, Penguin.

Harrison, A. and Jackman, R. 'Rate Support Grant', *CES Review*, 4, 22–6.

Harrison, A., Jackman, R. and Papadachi, J. (1979) 'Needs grant; which way now?', *CES Review*, 6, 18–31.

Haynes, G. and Bentham, R. (1979) *Community Hospitals and Rural Accessibility*, Farnborough, Saxon House.

Jackman (1979) 'London's need grant', *CES Review*, 5, 28–34.

Jackman, R. and Sellars, M. (1977) 'The distribution of the RSG: the hows and whys of the new needs formula', *CES Review*, 1, 19–30.

Johnston, R. J. (1978) 'Single-member European constituencies for London', *Representation*, 18 (72), 23–6.

Kirby, A. M. (1979b) *Education, Health and Housing*, Farnborough, Saxon House.

Kirby, A. M. (1982a) 'The external relations of the local state', in Cox K. R. and Johnston, R. J. (eds) *Conflict, Politics and the Urban Scene*, London, Longmans.

Kirby, A. M. (1982b) 'Analysis of variance in a descriptive context: a geographic example', *Bulletin in Applied Statistics*, in press.

Lambeth Inner Area Study (1977) *Inner London: Policies for Dispersal and Balance*, London, HMSO.

Rubin, B. (1979) 'Aesthetic ideology and urban design', Association of American Geographers, *Annals*, 69(3), 339–361.

Taylor, P. J. (1977) *Quantitative Methods in Human Geography*, Boston, Houghton Mifflin.

Taylor, P. J. (1979b) 'Difficult to let, difficult to live in and sometimes difficult to get out of', *Environment and Planning*, A 11, 1305–20.

Taylor, P. J. (1982) 'The changing political map' in Johnston R. J. and Doornkamp, J. C. (eds) *The Changing Geography of the United Kingdom*, London, Methuen.

Taylor, P. J. and Johnston, R. J. (1979) *Geography of Elections*, Harmondsworth, Penguin.

Zoloth, B. (1976) 'Alternative measures of school segregation', *Land Economics*, 52(3), 278–98.

8 Implications

Is this how they teach you at Oxford now? One reads last chapters first?
(John Fowles: *The Magus*)

As I outlined in the Introduction, this book is a personal attempt to justify the existence of geography as a discipline that examines phenomena from a spatial perspective. It is not a research agenda, nor even – as I shall show below – a particularly full account of the material under investigation.

One obvious drawback of the approach attempted here is that it is self-fulfilling. Clearly, any topic can be examined from a spatial perspective; it remains to be seen whether that perspective has any meaning, and the first part of this chapter attempts that review. The same can be said of the other main theme developed, which is a focus upon consumption issues. Once more, similar remarks are in order; it is possible to divorce a study of social relations entirely from issues of production, and to emphasize consumption – be it of housing or health care, or transportation or whatever. It remains to be seen, however, whether such an emphasis is feasible, or whether it is really rather like studying biology without taking any interest in where the birds and the bees actually come from.

In essence then, this chapter is an attempt at self-criticism, an exercise in pinpointing the weaknesses within the book as a whole. Let us begin, once more, with space.

Space – yet again

It is clearly right that there exists great scepticism about the efficacy of a spatial perspective; it makes little sense to attempt to translate every discipline into its geographical equivalent (geolinguistics? religeography?). Conversely, and as I have attempted to argue here at some length, there are instances where it is difficult to comprehend a process without viewing it in a spatial dimension; the malapportionment that exists in elections is a case in point.

The problem is to identify more rigorously what we mean by

phrases such as 'a spatial perspective', 'a spatial dimension' and 'a geographical approach'. Are they not hiding a rather loose conception of what spatial inquiry is about? Let me illustrate this with two examples.

Chaffinch song

A recent study by Slater and Ince focuses upon the ways in which bird songs evolve, and particularly upon the existence of geographical variations. However, as they note, these variations are not necessarily of interest in themselves: 'spatial variation in song has been studied in many species, but it remains questionable whether it is in itself an important phenomenon or simply a by-product of the role played by learning in ontogeny' (Slater and Ince, 1979). Put more simply, does the fact that bird song varies from place to place have anything to do with territoriality, or is it simply a result of poor learning behaviour from generation to generation?

The analysis suggests that some geographical variation certainly exists; as Table 8.1 indicates, some types of call are restricted to particular islands. This kind of information could imply that

Table 8.1 Geographical distribution of song types in the Orkneys; note that each chaffinch has more than one repertoire

Song type	Finstown (16 birds)	Balfour (15 birds)	Trumland and Woodwick (6 birds)
1	1	0	0
2	1	0	0
3	1	0	0
4	1	0	0
5	4	0	0
6	6	1	0
7	7	1	2
8	10	4	0
9	9	4	3
10	1	9	0
11	1	9	0
12	0	1	0
13	0	1	0
14	0	1	0
15	0	0	3
16	0	0	2

Source: Slater and Ince (1979) Table 1.

chaffinch song varies from location to location, with the process of learning contributing to the creation of a specific 'kinship group'. As it is, this seems unlikely due to the existence of multiple songs within any particular locality, and consequently it is assumed that distinct geographical patterns occur more as a result of mistakes that emerge as offspring learn from their parents; these mistakes then become features of a particular locality.

In the end the role of space is played down by the zoologists undertaking the research; their work shows clearly, however, a context within which a spatial perspective can be very simply articulated. Of course, this study possesses one advantage, in so far as the spatial domain in question is a natural one; as we shall see below, the man-made environment cannot be so simply treated.

Juvenile referrals

A recent investigation of juvenile crime and detection reveals that a number of factors determine whether a suspect is charged immediately or referred to the Juvenile Bureau, which may simply caution the offender, release him or her, or charge him or her. The type of offence is important, as too are the number of previous convictions or referrals. Age is important also, but most interestingly – and controversially – both race and location seem significant. All these variables are included in Table 8.2, which shows the probabilities of being referred to the Juvenile Bureau for a selection of offenders and three types of crime. The data are based on information collected on 1444 individuals arrested in 1978 within the Metropolitan Police area; the probabilities are based on the results of logit analysis (Wrigley, 1976).

Table 8.2 has been designed to focus upon the areal differences, and there are clear variations between the three units used. Area 1 comprises two inner-London boroughs, south of the Thames; Area 2 comprises two inner-London boroughs north of the Thames, whilst Area 3 comprises two boroughs in outer London. In relation to the three crimes examined, it can be seen that

(i) those arrested in Area 1 for violence are far less likely to be referred than those in Area 3;
(ii) those arrested in Area 3 for burglary are far less likely to be referred than those in Area 1;
(iii) those arrested in Area 1 for public disorder are far less likely to be referred than those in Area 3.

These results might be taken to imply some spatial bias in the operation of the referral procedure; an alternative explanation is at

Table 8.2 Areal differences in criminal referrals, London

Crime	Profile	Age	Ethnic group	Previous referral (R) and conviction (C)		Area probability			Difference between areas 1 and 3
				R	C	1	2	3	
Violence	1	15–16	White	1(+)	1(+)	0.328	0.357	0.619	0.291
	2	10–14	Black	1(+)	1(+	0.461	0.492	0.740	0.279
	3	15–16	Black	0	0	0.460	0.492	0.739	0.279
	4	15–16	Black	1(+)	1(+)	0.251	0.276	0.527	0.276
	5	15–16	White	0	0	0.542	0.585	0.805	0.263
	6	10–14	White	1(+)	1(+)	0.554	0.586	0.805	0.251
									Difference between areas 1 and 3
Burglary	1	10–14	Black	1(+)	1(+)	0.679	0.399	0.364	0.315
	2	15–16	White	1(+)	1(+)	0.625	0.343	0.311	0.314
	3	15–16	Black	1(+)	0	0.784	0.533	0.497	0.287
	4	10–14	White	1(+)	1(+)	0.809	0.560	0.534	0.275
	5	15–16	Black	1(+)	1(+)	0.454	0.207	0.183	0.271
	6	15–16	White	1(+)	0	0.879	0.696	0.664	0.215
									Difference between areas 1 and 3
Public disorder	1	15–16	Black	1(+)	1(+)	0.229	0.473	0.664	0.435
	2	10–14	Black	1(+)	1(+)	0.431	0.696	0.835	0.404
	3	10–14	White	1(+)	1(+)	0.459	0.720	0.850	0.391
	4	15–16	Black	1(+)	0	0.565	0.797	0.897	0.332
	6	10–14	White	1(+)	1(+)	0.685	0.867	0.935	0.251

Source: extracted from Landau (1981), Table 6.

hand, however. More simply, the results show the seriousness with which particular crimes are regarded by the police. Burglaries in Area 3 (a suburban location) constitute crime against property, by definition (it would seem) a more serious activity when carried out against the rich than when undertaken against the poor (in Areas 1 or 2). Conversely, violence in the streets is treated harshly in the inner areas. A clue to the reasons for this is found in the public-disorder category, for here are included arrests of 'suspected persons', i.e. those arrested for 'sus'. Landau suggests that 90 per cent of those arrested for 'sus' were taken in Area 1 – and in excess of three-quarters of these subjects were black. Very simply, it appears that this *spatial* affect is in fact a *racial* effect, and one that fits in entirely with the rest of Landau's analysis, which shows that 'black juveniles seem to be treated by the police more severely than their white counterparts' (1981, p. 42). (It also appears to corroborate the allegations made at the time of the Brixton riots in 1981 that undue police attention is paid to young blacks. Brixton is within Area 1).

The point of contrast between these two small examples should be clear. In the first instance, it is feasible to investigate a spatial process (albeit absent). In the second, the spatial effects are entirely secondary, the result of social segregation that places rich homes in one location (*vis-à-vis* burglary) and blacks in another (*vis-à-vis* 'sus). To induce from the data in Table 8.2 that there is some relationship between space and social processes would be dangerous: it could easily lead to the naïve assumption that there is some difference between police behaviour and attitudes north and south of the river, and that the spatial units in question are more than convenient bureaucratic divisions.

The criminological example underlines, then, the problems of reifying space: in other words of treating it as a thing-in-itself, of assuming that what goes on in Area 1 is necessarily different from what occurs in Area 3, simply because the data are different. This is a fundamental point, well exemplified by a recent debate between Peet and Smith.

'Death, degeneracy and radical eclecticism'

The Peet–Smith exchange is an interesting one, in so far as it focuses upon the relationships postulated in Chapter 1 to exist between spatial relations and social processes. Peet's stance is that economic and social progress in any society is moulded by the fact that there exist other societies located elsewhere within the world economy; as a result, changes may always occur as a consequence of, for example, technical innovations transferring from one national economy to another (Peet, 1981, p. 109). This type of process he identifies as 'spatial dialectics'.

In one sense, Peet's argument is philosophically not really very far removed from the one expressed in the previous section, i.e. because there exist spatial variations within phenomena, it is fair to assume that a process related in some way to the different locations has been identified; more simply, if society were to be collapsed into one location, then logically these variations should also disappear. As Smith points out, however, space is again being reified. To take Peet's example, the spatial role is wrongly expressed:

the production of space under capitalism is precisely the uneven development of capitalism. It is the development of underdevelopment which has been observed to take place at different spatial scales. This way of putting it, of course, is in accord with the old metaphysical view of space. It is not underdevelopment which *takes place* at different spatial scales, but rather the processes of development and underdevelopment that produce these different spatial scales as identifiable and differentiated. (Smith, 1981, p. 117, original emphasis)

Let us not, however, enter this aspect of the debate too deeply. The important point is this: to what extent can Smith's argument be transferred to the kind of material analysed in the earlier chapters? Are regional health authorities or school catchment areas totally artificial and meaningless spatial partitions, or do they stem directly from the organization of society? If they are the product of mere bureaucratic chance, then it serves us very little to reify them, and then to examine the deprivation that they foster. If, however, their creation is an inherent part of social organization, then such an approach has some validity: we can in other words insert 'jurisdictional boundaries' into Smith's statement, instead of 'spatial scales'.

The problem here is that we enter a relatively unresearched area. Clearly, the manipulation of space by capital interests involves fewer considerations than the juridical manipulation of space, which involves the state, local organization, political struggles, capital needs and so on. This 'institutional organisation of space' (Gregory, 1978, p. 118) is seen by some as possessing a logic; Castells, for instance, writes of '"urban planning" [dealing] with the problems of the functioning of the ensemble by cutting them up into significant spacial [sic] units based on networks of interdependencies of the productive system' (Castells, 1978, p. 27). The ways in which this logic evolves have, however, not yet been examined in detail, and the perils of 'radical eclecticism', as Smith terms it, are real ones.

This must be the major limitation of this book. The validity of using given spatial units (LEAs, RHAs, constituencies) has not been proved; nor has the evolution of these units and their boundaries been explored. Brief reference has occasionally been made to the state, the local state, managers, political struggles, etc., but in this context the issue of explanation, rather than description, has barely begun.

Consumption, production and political action

A further issue relates to the possibility of examining distributional questions in isolation, as has tended to be the case here. In Chapter 2 I recognized the existence of social status as a dimension that may transcend social class, using the latter in the sense of income groups; in Chapters 5 and 6 I examined the existence of political cleavages arising from distributional (externality) issues. Clearly, both these concerns are difficult – if not dangerous – when abstracted from other issues. To take the simplest level of argument, it is not enough to suggest that consumption simply 'is'; it is, of course, a social process, in so far as some aspects of consumption are organized by the state for particular purposes (see, for example, Saunders, 1980, Chapter 3). It is thus organized for the benefit of particular groups and locations, and is part of the process of capital accumulation and circulation. My focus has, though, been only upon the spatial implications of consumption, and this has, at this stage, involved an incomplete consideration of these relationships.

Effectively, the same sorts of comments are in order *vis-à-vis* the identification of political cleavages arising from spatial questions. As Saunders once more points out, such a perspective involves the recognition of a 'necessary non-correspondence between economic class categories and political action' (1980, p. 100). This is in itself not a problem, as long as we can be sure that spatial issues are just that, and not simply extensions of other cleavages – perhaps relating to the ownership of land. There is sufficient evidence to suggest that a spatial cleavage *can* occur, and the material in Chapters 5 and 6 illustrates this. What is, however, more difficult to interpret, is the importance that can be attached to these ephemeral political struggles. Do they constitute 'new' tensions, in the sense that Dunleavy has examined the growth of consumption cleavages? Have they occurred at particular historical moments? Do they reflect some process of politicization, on the road to a greater involvement in conflicts of other types? Or do they constitute some form of false conflicts, encouraged – like public inquiries themselves – as evidence of the open and participative nature of society? Once again, however, there are no available answers.

Conclusions

In one sense then, this book ought not to conclude at all, in so far as it has simply begun a task of description; and the task of explanation should now begin – preferably over the page. The perceptive reader will have noted, of course, that it does not, because this author – for one – is still unsure as to how the task of explanation is to be completed. It is tempting to argue that spatial issues (more particularly, the existence of boundaries and the organization of society into distinct geographical units) can in some way influence social change; as the Peet–Smith debate illustrates, this may, however, artificially promote the importance of these questions. The logical outcome of that debate is to relegate the role of geographical research to some form of description, which can then be utilized, by others, within broad social-science enquiry. For Gregory this process has already occurred:

it is therefore high time to abandon the pretence of a separate existence for geography, one which has reduced it to a repository of low-level propositions which are at best self-evident and at worst simply false (1978, p. 171).

Only one point is clear: as geographers begin to peer cautiously about them, to ask what other modes of enquiry have achieved, rather than bleating plaintively about why they are not taken too seriously by others, then they realize just how little they do at present have to offer. This cannot continue. We cannot accept that human geography (which is now the only branch of the subject that still maintains any research momentum) can continue 'branching towards anarchy' (Johnston, 1979a, p. 189). Even the punks saw the danger in that.

References

Castells, M. (1978) *City, Class and Power*, London, Macmillan.
Gregory, D. (1978) *Ideology, Science and Human Geography*, London, Hutchinson.
Johnston, R. J. (1979a) *Geography and Geographers*, London, Arnold.
Landau, S. F. (1981) 'Juveniles and the police', *British Journal of Criminology*. 21(1), 27–46.

Peet, R. (1981) 'Spatial dialectics and Marxist geography', *Progress in Human Geography*, 5(1), 105–10.

Saunders, P. (1980) *Urban Politics*, Harmondsworth, Penguin Books.

Slater, P. J. B. and Ince, S. A. (1979) 'Cultural evolution in Chaffinch song', *Behaviour*, 71, (1–2), 146–66.

Smith, N. (1981) 'Degeneracy in theory and practice: spatial interactionism and radical eclecticism', *Progress in Human Geography*, 5(1), 111–18.

Wrigley, N. (1976) 'An introduction to the use of logit models in geography', *CATMOG*, 10, Norwich, Geo-Abstracts.

Name index

Subject index